THE ART OF MATHEMATICAL

PROBLEM SOLVING

THE ART OF MATHEMATICAL PROBLEM SOLVING

RICHARD M. BEEKMAN

Copyright

Beekman, Richard M., 1958-

The Art of Mathematical Problem Solving

Printed in the United States of America

First printing, 2016

This book is dedicated to Sebastian, the little genius. I probably have more to learn from him than he can learn from me.

About the Author

Richard M. Beekman is an electrical engineer and mathematician. Mr. Beekman specializes in number-theoretic combinatorics, and he has written over 100 research papers in that specialty. He is the inventor of the *logarithmic generating function* and the *obverse generating function* which are new tools in the field of number-theoretic combinatorics. On www.brilliant.org, under the pseudonym of "Igmut Schnoll," he is rated at the highest possible level, *Level 5*, in Algebra, Geometry, Number Theory, and Combinatorics. Mr. Beekman has been studying mathematics for over 50 years. He lives in St. Louis, Missouri, along with numerous uninvited arthropods.

Preface

Many Olympiad caliber mathematics problems can be easily solved if you have the right clever idea. The difficulty is that the "right clever idea" is often so clever that it looks like a magician pulled a rabbit out of a hat. When you see the clever idea, you feel stupid, because the idea is just too ingenious and too far afield from our normal thought processes. You feel like you could not possibly have thought of such an idea by yourself. Surely, only a super-genius could have such insights. This poses a serious challenge for the mathematical problem solver. If you are not a natural born super-genius, does that mean that there is no hope that you can ever solve a really hard mathematics problem? Is there some way that mortal human beings can break through the mental obstacles and discover the right clever idea that solves the problem?

After many years of mathematical problem solving, I have some insights into the process of solving difficult mathematics problems. First, I challenge the notion that mathematics is the sole domain of natural born geniuses. The truth is that humans are just not that good at mathematics. Human brains are not designed for doing mathematics. It is an evolutionary side-effect—an accident of nature—that we can do any mathematical thinking at all. I believe that mathematical genius is more the product of hard work than of any natural born inclination.

Second, the seemingly miraculous solutions to mathematics problems are generally the result of intensive exploration. The appearance that a super-genius drank a cup of coffee on a Sunday afternoon and leisurely conceived the right brilliant idea is mostly an illusion. What you generally don't see, when you read the solution to a published mathematics problem, is all the hard work and failed attempts that went into discovering the solution.

The key to solving many difficult mathematics problems lies in fundamental patterns that are revealed by the problem itself. After reading and understanding a problem, you must explore the problem. The exploratory process produces data, and that data reveals patterns. Patterns are almost always important, and it is precisely those patterns that provide the key idea—the brilliant insight—that allows us to solve the problem. Once you identify the pattern, you state it as a conjecture. Then you prove the conjecture using standard mathematical proof techniques. At that point, your solution is close at hand. You have found the key idea to solve your problem. From this perspective, the essential important skill in mathematical problem solving is the process of exploring the problem. It is *exploration*, not mythical genius, that is the central theme of successful mathematical problem solving.

In this book we will focus on the process of mathematical exploration. You may not be able to solve every mathematics problem that comes your way, but you will understand the *process*. And you will understand how mathematicians approach novel mathematics problems. Learning how to think critically and creatively are important learned skills in the fine art of mathematics.

Richard M. Beekman

St. Louis, Missouri

September 2015

Contents

Acknowledgements

I am deeply indebted to ancient mathematicians, such as Archimedes (mechanics), Euclid (geometry), and Apollonius (conic sections), who pioneered disciplined thinking in mathematics. Archimedes, especially, taught me to think deeply about simple ideas. I also extend my gratitude and sincere appreciation to all the great modern Russian mathematicians, such as Andrei Kolmogorov, whose expository writings are truly outstanding. Without the Russians, and their great books on mathematics, I would have suffered from intellectual poverty.

I am greater than or equally grateful, with equality if proportionality applies, to Brilliant (www.brilliant.org). Brilliant is a magnificent organization, and their website has provided me with countless wonderful challenges and motivations in mathematics. Many of the problems in this book were posted on Brilliant, though all of the solutions are my own to explain my thinking process. My goal is to teach mathematics as a fine art.

My dear friend and colleague, Christian Hasselberg, deserves special mention. He is a creative and original thinker, especially in business and management. He has always had an iconoclastic, playful, and innovative temperament which further inspires me to study mathematics. Unlike many people, Christian never asked me the question, "Why do you waste so much time studying mathematics?" Instead, his attitude has been, "What else would you do with your time?"

The problems in this book came from various sources. Some of them are mathematical classics—part of the mathematical folklore. Others came from various competitions, journals, or were posted on www.brilliant.org. Problem sources, if known, are presented at the end of the book, before the index, in the section entitled "Problem Sources."

Introduction

There is a wonderful website for people who enjoy solving challenging mathematics and physics problems. It is www.brilliant.org. The problems posted on this website are not routine exercises. They are real mathematics and physics problems, often of research or Olympiad caliber, that typically require considerable ingenuity and hard work to solve.

I began my love affair with mathematics at the age of six. I remember looking at an advanced mathematics book, with all its strange symbols and incredible facts, and I thought, "How can people know such things?" Since that time, I have studied mathematics for 30,000 labor hours over half a century. I have read hundreds of mathematics books (I stopped counting at 231), and I have invented or discovered dozens of new theorems in number-theoretic combinatorics, which is my research specialty. On brilliant.org, under the pseudonym of "Igmut Schnoll," I am rated at Level 5, the highest possible level, in four categories: algebra, geometry, number theory, and combinatorics.

One day, a friend of mine, who is rated at Level 3 on brilliant.org, brought me a mathematics problem that he had been struggling with all night. He needed help, so he asked me if I could solve the problem. I quickly found two different solutions in about 20 minutes. My astonished friend asked, "How did you do that?" My response was, "Honestly, I haven't got a clue how I did that." After you have been studying mathematics, or any subject, for half a century, things start to gel in your brain. You finally start to "get it." Your brain automatically sees patterns and trends that are quickly exploited. It's a magical feeling. Still, my friend's question haunted me. It is, after all, a valid question. How do we actually solve a difficult mathematics problem? What process should we follow to achieve success? I started thinking deeply about the actual problem solving process that I tend to follow. There is, indeed, a method to the madness. I decided to document that process both for my benefit and for other hopeful mathematical problem solvers.

My goal is to emphasize the *art*, rather than the *science*, of mathematical problem solving. Art suggests passion, feeling, and maybe a little vodka. People who do not understand mathematics, especially the fine art of mathematics, often think that the subject is about applying cold-hearted, reptilian logic to sterile quantitative exercises. Logic certainly has a role to play in mathematics. However, for serious mathematical problem solvers and research mathematicians, intuition, passion, and heartfelt feelings dominate the exploratory process.

Rather than presenting a collection of theorems, followed by several exercises, I want to avoid a tool-centered mentality. Theorems, while important, are like tools that must be applied judiciously. Tools do not make the master carpenter. It is the master carpenter who makes good use of tools. For this reason, theorems will be introduced sparingly, as needed, within the context of solving mathematical problems. The *process* of exploring a mathematics problem is far more important than rote memorization of theorems. We want to learn how to proceed from a state of panic and fear, when we first read a mathematics problem statement, through a meandering exploratory process that eventually leads to a beautiful and ingenious solution. I'm going to show you how to do that. Logic alone will not get you there. You need passion, intuition, imagination, courage, irreverence, and persistence.

I should mention that there are many different, though similar, problem solving processes. My process is not unique, nor is it entirely original. It is simply the process that works for me. No two problem solvers follow exactly the same process. People have different aptitudes, preferences, and personalities. Yet, there are some unifying common ideas, concepts, principles, and methods that can be studied, learned, and mastered. My intent is to present my problem solving strategy and process with the hope that you can learn from this process and adapt it to your own creative style. After that, all you need is lots of practice. Practice, more than innate ability or intelligence, is the secret to success. Brilliant mathematicians are not born; they are forged, like steel, in the cauldron of solving challenging mathematics problems. I wish you the best of luck in your endeavor to become a world-class mathematics problem solver.

The book is divided into two parts. Part I discusses the process of mathematical problem solving. Part II contains difficult mathematics problems with full discussions and solutions. The problems in Part II are generally Olympiad caliber mathematics problems. It's like the Olympics for mathematicians. But don't worry. I want to create the feeling that you are sitting around the fireplace solving mathematics problems with your dear Uncle Rich. We will talk ourselves through the solution to a difficult problem by engaging our ideas, feelings, intuitions, hunches, and playful irreverence. This is how real mathematicians work, and it is, perhaps, why we tend to be an eccentric lot of fellows.

Part I

THE PROCESS OF MATHEMATICAL PROBLEM SOLVING

Chapter 1

The Problem Solving Process

The chapters in this book are organized around a general mathematics problem solving process. With the exception of this first chapter, each chapter discusses one step in the overall mathematics problem solving process.

To be a good mathematics problem solver, you must be willing to venture into the unknown, take bold risks, explore new territory, and possibly get lost along the way. My general problem solving process can be listed as a sequence of steps, but you should understand that, in actual practice, the steps are not always executed sequentially. When your ideas fail, and you get lost, you may need to backtrack or go back to earlier steps in the process. Don't worry about that. It is entirely natural. It is part of the problem solving process to make mistakes and start over again with new ideas.

Here are the basic process steps for solving difficult mathematics problems:

Step 1. Overcome your fear. (This is the psychology of problem solving.)

Step 2. Understand the problem.

Step 3. Explore the problem and look for patterns.

Step 4. Formulate conjectures.

Step 5. Prove your conjectures and solve the problem.

Step 6. Verify your solution.

Step 7. Polish the stone.

Step 8. Reflect and learn.

The eight process steps comprise the process that I generally follow to solve some really hard Olympiad caliber mathematics problems. We just need to put some beef on the bone. We will do that in Part II of the book, when we sit by the fireplace and talk through the solutions to some good mathematics problems.

When we solve mathematics problems, we must be flexible in our thinking. We should think of the eight process steps as a general *framework* for solving a novel mathematics problem, and not necessarily as a prescription that must be followed in every case.

Before we move forward into the details of each of these problem solving process steps, let us discuss, for a moment, what a mathematician means by "solving a mathematics problem." First, mathematicians are interested in *problems*, not *exercises*. Exercises are the things taught in public schools. They are standard, typical, vanilla, off-the-shelf problems of a certain well-defined and well-understood type. The "solution" to these exercises amounts to little more than plugging some numbers into a magic formula, such as the quadratic formula, that was apparently bequeathed to us by the gods on Mount Olympus. You grab your calculator, mindlessly plug in some numbers— thinking is optional—punch out the answer, and get an "A" in your high school algebra class. Trust me, that is *not* what mathematicians mean by a "problem." A real mathematics *problem* is something new, at first frightening, where you feel lost, possibly hopeless, without a compass, without a roadmap, and without a magic elixir from the gods. You are all alone on a dark road, in a dark forest, on a dark night. You, and you alone, must solve this problem. It will require you to dig deep within yourself, brave new ideas, risk failure, and try again. A real mathematics problem is something where you are initially lost and hopeless, and you must invent your way through it, possibly even inventing new mathematics and proving new theorems along the way.

A *solution* to a mathematics problem is *not* just an answer. It is not simply a number, like 5. When a mathematician asks for a "solution" to a mathematics problem, she is asking for a well-reasoned, thoughtful, insightful, and clearly articulated sequence of logical steps that demonstrate, with unequivocal certainty, that a certain mathematical statement or conclusion is true. This is a very high standard of excellence in critical and creative thinking.

Let us now explore each of the steps in the mathematical problem solving process.

Chapter 2

Overcome Your Fear

The psychology of problem solving

The most difficult part of mathematical problem solving is getting past the initial psychological shock of the problem statement. Mathematical problems are usually stated using special symbols with technical jargon, and they can be quite intimidating to people who are not familiar with the language of mathematics. Even if you are familiar with the language of mathematics, most real mathematics problems—research or Olympiad caliber problems—do not resemble anything that you have ever seen before. Real mathematics problems, as opposed to exercises or routine drills, are difficult even for the best mathematical minds. Humans are just not that good at mathematics. Our natural, untrained, abstract thinking abilities are just barely adequate to comprehend mathematics on any level. Even the most brilliant mathematical minds must struggle, often for many years, to comprehend difficult mathematics problems. Any thought that you are stupid, inadequate, or not smart enough to solve mathematics problems will only hurt you. You must get any thoughts of inadequacy completely out of your head.

Practice, practice, and practice some more.

What does it really take to become a good mathematics problem solver? You might think that the answer is something like natural ability, or aptitude, or having a high IQ. Studies have actually been done on this subject. It turns out that the primary factor in becoming an expert at anything is *practice*. Generally, it takes about 10,000 hours of study and practice to become an expert at anything—playing the violin, nuclear physics, or mathematics. This is good news. It means that you have the potential to become a great mathematics problem solver. All you need to do is study and practice. That's it!

Think like you are already a great problem solver.

It is also important to have the right attitude when you solve mathematics problems. Instead of thinking, "I'm not good at mathematics, so I could never solve a problem like this," you should instead think, "This is an interesting problem. Maybe I can discover and prove a new mathematical theorem."

To me, thinking that you cannot possibly solve a mathematics problem is like a fisherman thinking that he cannot possibly catch a fish in the ocean. Why would you think such a thing? I suspect that part of the answer is that most of us have never solved a real mathematics problem. We were beaten into a stupor by a school system that is not designed to produce independent and creative thinkers.

Mathematical myths

We must break old myths that no longer serve us and replace them with a better mental model of mathematical problem solving. Here are a few myths that you may have heard. These myths are so absurd that I will merely list them.

Myth 1. You must be a genius to understand mathematics.
Myth 2. There is nothing new to discover in mathematics.
Myth 3. You need a Ph.D. to discover new mathematics.
Myth 4. Only men are good at mathematics, not women.
Myth 5. Only young people produce great new ideas.
Myth 6. Mathematics is nothing more than logic.
Myth 7. Feelings have no place in mathematics.

Don't try to be brilliant.

Myth 1 is an especially debilitating myth for anyone who wants to become a really good mathematical problem solver.

There is a popular misconception that you must be brilliant to understand mathematics, and you must be a genius to invent and discover new mathematics. This is nonsense! Genius is a peculiar concept. Some people think that you are born a genius. Others think that genius is the product of divine inspiration, as if Zeus strikes you with a bolt of lightning at a decisive moment. The problem with these views of "genius" is that they are unreliable. You can't run a business on these notions. I have an alternative view of genius: *Genius is the result of hard work.*

There are mathematical problems that look hard but have simple, clever solutions. If you approach a problem thinking that you are going to be a genius and find a brilliant solution, you run the risk that you will waste a lot of time on wild, "outside-the-box" attempts and discover nothing at all. A better approach is to focus on the task at hand. Don't try to be brilliant! Instead, explore the problem. Break the problem apart, look at small cases, collect numerical data, analyze the data, and look for patterns. After some hard work, you will find patterns. Those patterns will eventually lead you to the right solution—and that solution may turn out, after all, to be simple and brilliant.

Be persistent.

Only the easiest, most routine exercises will yield to your initial attempts to solve the problem. Real mathematics problems are difficult, and they will challenge your creative abilities to find a solution. Persistence definitely pays off in the long run. A good problem solver is relentless. You must develop the mental capability of refusing to give up on your problem. If you absolutely cannot solve your problem, you may need to tuck it away in the back of your mind for a while and come back to it later. Try working on some other problems. Eventually, an idea will pop into your head that allows you to solve your problem. Don't give up the fight!

Mathematics is a fine art.

I hope to show you, especially when we solve problems in Part II, that mathematics is a fine art. Like all fine arts, it involves human passions, feelings, and intuition. Anybody can do great mathematics—beginners and experts, men and women, even the guy at the bus stop. Your math anxiety will disappear when you learn to see mathematics as a medium for open-ended intellectual expression. Mathematical knowledge is not carved in stone. It is always changing and evolving, and there is plenty of room for *your* new ideas. Anything goes.

Mathematics is a beautiful subject. It is a fine art like poetry, painting, music, dance, and sculpture, although its disciplinary constraints are intellectually demanding. Its masterpieces are abstract constructs of the human mind that reflect the underlying elegance and simplicity of the cosmos. Do not fear mathematics. Learn to create and invent your own mathematics.

Turn failure into success.

What happens if you try really hard to solve a problem and you fail to solve it? Generally, what happens is that you work really hard on a problem and you end up solving something else. You always end up solving something. The thing you end up solving may not be the problem that you wanted to solve. However, in trying to solve the given problem, you probably have explored, calculated, and discovered something interesting. In that case, turn your failure into a success by turning your discovery into a new theorem. Whatever you have discovered by accident, as long as it is interesting, qualifies as a new theorem. This is often how professional research proceeds. We try to solve one problem—which we fail to solve—and we end up solving another problem. That's OK, and it's a good thing. Write a research paper. Describe your new problem (the one that you accidentally solved), and pretend like it was the problem that you intended to solve all along. Present your theorem and your beautiful and brilliant proof, which you already know (because you stumbled on it accidentally), and nobody will know better. Turn your failures into successes!

Chapter 3

Understand the Problem

The most important step in the mathematical problem solving process is to understand the problem that you are trying to solve. This step is more important than all the other problem solving process steps because it is the place where people tend to falter and fail. It's human nature that we want to hurry up and get on to the fun stuff—exploring and solving the problem. But if you don't clearly understand the problem, its constraints, and its objective, you will either fail to solve the problem or you will solve the wrong problem. I have done that often enough that I now read each problem *three times* before I even think about exploring it. It's embarrassingly frustrating to invent a brilliant solution to the wrong problem. Don't do that!

Suppose, for example, that a problem asks you to find the sum of roots of a polynomial. If you don't pay close attention to the problem statement, you might go off on a wild goose chase trying to calculate all the roots. But the problem did not ask you to find the *roots* of the polynomial. The problem asked you to find the *sum* of the roots. That is a different problem, and there are shortcuts that we can use that allow us to avoid finding the actual roots. We could use Vieta's algebraic relations to easily find the sum of the roots from the coefficients in the polynomial. In other words, we can find the sum of the roots without ever finding the roots themselves.

The process step of "understand the problem" is actually a process of its own. And it's a process that you really don't want to cut corners on. You want to be exceptionally well-disciplined in the step of understanding the problem. What do we mean by "understand the problem?" How do we do it?

Read the problem statement three times.

The first thing you should do with a new mathematics problem is to read the problem statement at least three times carefully. You want to make sure that you really understand the problem statement. Learn how to do this actively rather than passively. When you *actively* read a problem statement, you jot down notes with pencil and paper. If the problem is stated using technical jargon, then try to restate the problem using your own words in everyday language. If there are equations or inequalities in the problem statement, try replacing some of the variables with numbers. Do a few calculations to see what happens. Anything that helps you to understand the problem is good. You cannot solve a problem that you don't understand. If the given problem appears too difficult for you, don't be afraid to simplify it. Try solving an easier version of the problem first.

That may help you understand how to approach a solution to the harder problem.

Draw a picture.

Whenever possible, sketch a picture or figure of your problem. This is obvious if you have a geometry problem. But it also works for algebra problems. Try to translate variables into numbers. Graph equations to see what they look like. If there are numbers in the problem, try to represent them visually as blocks stacked up in some pattern. The human mind is exceptionally good at visual-spatial reasoning. Our minds think visually, using pictures. So, whenever possible, draw a picture to illustrate the problem that you are trying to solve.

What kind of problem is it?

What kind of problem do you have? What branch of mathematics does it belong to? These are good questions to ask, because it will give you an idea of what methods or theorems apply to your problem. A geometry problem will require visual-spatial methods, constructing auxiliary figures, and, of course, well-known geometry theorems. If your problem looks like an algebra problem, then you will likely need to use some ideas and theorems from algebra, such as factoring polynomials, to solve the problem successfully. Being able to classify the problem and identify what kind of problem it is will significantly cut down the size of the problem space. It helps to focus your attention on those methods and theorems that are most applicable to your problem.

George Polya, a great mathematics teacher, suggested that there are two fundamentally distinct types of mathematics problems: "to find" and "to prove" problems. Problems that have the flavor of "to find" are asking you to find something. You might need to find a construction, an object with some property, or a number. A combinatorics problem, for example, might ask you to find the number of different arrangements of a set of objects. Or, maybe the problem asks you to find the maximum value of some function over some domain. These are "to find" problems. There are also "to prove" problems. Problems that ask you to prove something typically give you a mathematical statement, such as "the square root of 2 is irrational," and you are asked to prove the given statement. Sometimes, you are asked to prove that a statement is false. It is helpful to ask yourself these kinds of questions, because they help to focus your mind on the right kinds of ideas that you will need to solve the problem.

List the constraints.

Look carefully at the constraints, if any, for the problem. The problem might say that "x is a positive integer." Or it might say that "x is a nonnegative integer." It's easy to miss the subtle distinction between these two statements, but the distinction is important. The two statements, while similar, imply two different things. If x is a positive integer, then it is a whole number that cannot be negative or zero. However, if x is a nonnegative integer, then x could possibly be zero. Sometimes these apparent trivialities make a huge difference when you try to solve the problem. In mathematics, small details are important. Don't be afraid of constraints. Constraints are often your best friend. Constraints often restrict the kinds of methods and theorems that apply to your problem. You should learn to love them. And always write them down on paper. Writing things down is part of the problem solving discipline.

List your assumptions.

Write down your assumptions. Sometimes you will need to make some assumptions in order to make progress with your problem. This could happen either because the person who wrote the problem statement forgot to mention some important detail, or you may simply find it advantageous to make some assumptions to move forward with your solution. In the latter case, you will need to verify your assumptions later in the problem solving process. An example, which I actually encountered, involves the "square root function." Ask anybody what the square root of 4 is, and they will tell you "2." But −2 is also a square root of 4, because $-2 \times -2 = 4$. So, if you see a square root in your problem, does that mean the "positive square root function," which is restricted to the positive roots, or can we also have negative roots? If you don't know the answer, you may need to list it as an assumption and move forward. Sometimes the best strategy in mathematical problem solving is to just believe in a benevolent math god and move forward despite all obstacles.

What does a solution look like?

Ask yourself what a "solution" looks like. This may sound like a silly question, but it's not. In my specialty, enumerative combinatorics, a solution to a problem could be a number, a generating function, a recurrence relation, an asymptotic relation, an algorithm, or an explicit formula. Depending on the nature of the problem, any one of those things might be considered a valid solution to the problem. So, ask yourself, "If I successfully solve this problem, what will my solution look like?" Does the problem ask you to find a number, a proof, a figure, a formula, an algorithm, or what? What does success look like?

Chapter 4

Explore the Problem and Look for Patterns

In this chapter, we will discuss the basic principles of exploring a problem to find patterns. The applications will be presented in Part II when we actually solve some hard mathematics problems. Exploring a problem and looking for patterns is the essence of the fine art of mathematics. This is what real mathematicians do while doing research. The act of thinking about mathematical patterns often consumes us, like an obsession, and our best ideas sometimes occur while taking a shower. An obsessive-compulsive mind—one that cannot let go of a problem—is an asset for a mathematician. As part of a job interview process, a psychologist once asked me, "Do you count the tiles on the bathroom floor?" My response was, "Of course I do! I'm a mathematician."

Identify your brute-force-dumb solution.

The universal algorithm for solving every mathematics problem that ever existed, or that ever will exist, is simply this: *Multiply by zero and add the answer.* That's the cheater's solution. Seriously, though, the cheating approach to solving a mathematics problem has some merits, and it deserves serious discussion. Sometimes you are just lost and hopeless. You don't have a clue how to solve the problem. In cases like this, a *brute-force-dumb* approach is sometimes useful.

What do I mean by "brute-force-dumb"? Suppose you have a difficult geometry problem that asks you to find some angle in a figure. The brute-force-dumb solution is to get a piece of graph paper, a pencil, and straightedge, and carefully draw an accurate figure on paper. Then take a protractor and measure the angle. When you do this, you may find that the answer is 45 degrees. You may say, "Ah, but that's cheating. You didn't use geometry theorems to *prove* that the angle is 45 degrees." That's true, but don't forget that you are engaged in the process step of exploring the problem. If you explore the problem and find that the angle is 45 degrees, that's quite a revelation. Why? In geometry, angles of 45 degrees are very special angles. This knowledge may help you focus your efforts on 45-45-90 triangles. You can look up some theorems about 45-45-90 triangles in a reference book.

Other examples of brute-force-dumb solutions are writing computer programs to solve a problem by number crunching, solving probability problems by rolling dice, graphing a function to find its roots, and building a physical model.

Identifying your brute-force-dumb solution serves three purposes. First, it provides psychological comfort knowing that you could actually solve the problem if you had to. Knowing this will help you relax and tap into your playful, creative mind.

Second, the brute-force-dumb solution may actually give you the answer—like a number—to your problem. It is often easier to solve a problem if you already know the answer. (See Problem 2 in Part II for an example of this idea.) Knowing the answer points your investigations in the right direction so that you don't waste effort on approaches that won't work. In the hypothetical geometry example mentioned previously, you would not waste time looking up theorems about 30-60-90 triangles if you know that the angle is 45 degrees. Finally, the brute-force-dumb solution often points the way toward a simple, elegant solution. The brute-force-dumb solution becomes a roadmap, and you find shortcuts to streamline your path on that roadmap. The end result, after some refinement, may be a nice solution.

Get an answer.

One of the most important problem solving principles that I often use is the idea that *it is easier to find a solution if you already know the answer*. As soon as you know that the answer to your problem is, say, 0, then you must realize that it's not an accident. The math god made it come out that way. In mathematics, anything that comes out to be zero is significant. We can forget about looking at hyperbolic cosine functions. There is no need wasting time with partial differential equations on a Hausdorff manifold. No. This is something more basic, and the problem probably has a simple, ingenious, beautiful solution. Your job is to find it.

Suppose that your ship became stranded on a deserted Pacific island. The captain tells you to go find some water. Where should you look? Should you climb into the highlands? Should you try digging a well? Now, suppose that the captain told you the answer: There is a lake on the island. That knowledge is a game-changer. If we know that there is a lake on the island, then we can use logic, reason, and geography to figure out where the lake most likely resides. It's not going to be on the side of a cliff or along a rocky shore. So get an answer to your problem, even if you must use brute-force-dumb methods. This is part of the investigative process.

Explore the problem.

Hard mathematics problems are rarely solved directly. You cannot achieve victory against a heavily fortified enemy unless you find a weakness in his defenses. Finding a weakness requires exploration. The solution to a problem is obvious only after you know what the solution looks like. If all you do is read a problem and then immediately look up the solution in the back of the book, you will never improve your problem solving skills. You must *learn the art of suffering*. You have to struggle with a problem and endure your hardships so that you will appreciate and learn from the solution when you see it. This is part of the learning process. Exploring a mathematics problem is to engage yourself wholeheartedly in a battle with your problem. It's not easy, but the rewards are great.

How can you get started in the process of exploring your problem? We can't reduce exploration to a simple set of rules, because every problem is unique. However, there are some general principles we can follow.

A good place to start exploring your problem is to *begin with small cases, special cases, and extreme cases*. If a problem asks you to add 1000 terms, begin by adding two or three terms, simplify the result, and see what happens. Do you see any pattern beginning to emerge?

Sometimes you can substitute small numbers, like 1, 2, or 3, into your problem and see what happens. Or, if your problem involves a large parameter, like $n = 1000$, explore a smaller case, like $n = 10$. Often, a pattern will appear that you can generalize to the larger case. Here is a tip when choosing small cases. *Try to choose a small number that has the same mathematical properties as the big number.* If, for example, $n = 425$, then try exploring the case of $n = 5$. Why? Here, both 425 and 5 are odd, and they are both multiples of 5. If a problem is stated using a large odd prime number, like $p = 35993$, then try using a small odd prime like $p = 3$.

Also, try looking at special cases. What happens if x is 0 or 1? Zero and one are important special numbers in mathematics, and you should always ask what happens when mathematical variables are zero or one. Try special kinds of numbers. Sometimes special numbers have special properties that are relevant to your problem. See what happens when your unknowns are prime numbers. What happens if you use even numbers or odd numbers?

Another approach is to look at extreme cases. What happens if some part of your problem is either very large or very small? What happens if a variable tends to plus or minus infinity? Many combinatorics problems have complicated exact solutions for small values of a parameter, say n, but they become very well behaved when n is large.

You will often gain considerable insight into how to solve your problem simply by looking at small cases, special cases, and extreme cases.

Another important exploratory technique is to *do some calculations with numbers*. This is especially helpful with problems in number theory and algebra.

For number theory problems, experimental calculations with numbers, possibly using a spreadsheet computer program, may help you find a few numerical solutions to your problem. Once you have a few solutions, you can look for patterns in the solutions. Suppose, for example, that the solutions you find appear to be square numbers like 1, 4, 9, or 16. You should immediately suspect that maybe all of the solutions are square numbers. This observation becomes the basis of a conjecture. You can then focus your efforts on proving that conjecture.

For problems in algebra, you may feel confused, even overwhelmed, by too many variables, like x, y, and z. Plugging some actual numbers into an expression may help you understand what the expression is telling you. You may gain some insight into why it works. Suppose you have some complicated expression $f(x, y, z) = 0$. You don't know how to begin exploring this expression algebraically. So you play around with different numerical values for x, y, and z. After some experimentation, you may find that whenever $x + y + z = 0$, like $-5 + 2 + 3 = 0$, you get $f(x, y, z) = 0$, or $f(-5, 2, 3) = 0$. In that case, you should suspect that $(x + y + z)$ is a factor of $f(x, y, z)$. This means that you should try factoring the expression $f(x, y, z)$ into a product of factors, one of which is $(x + y + z)$.

The exploratory process is about having fun, playing around, and trying different things. Make guesses, do some calculations with numbers, follow your intuition and hunches, and look for patterns in the data.

It helps to stand back once in a while and look at the big picture. When you are working on a problem, ask yourself if there are other ways of looking at the problem. Try changing your perspective. Shift gears. An algebra problem, for example, may have a geometric interpretation, or it may have a topological interpretation. Suppose you are asked to find solutions, ordered pairs of real numbers (x, y), satisfying the nonlinear system of equations given by

$$\begin{cases} 2x^2 - 6xy + 2y^2 + 43x + 43y - 174 = 0 \\ x^2 + y^2 + 5x + 5y - 30 = 0 \end{cases}$$

Viewing this as an algebra problem, it looks frightening. Viewing it as a geometry problem, you might realize that the first equation is an analytic equation of a hyperbola, while the second equation is an equation of a circle. Now the solution is at least conceptually simple. A brute-force-dumb approach is to simply graph the hyperbola and circle and read off the intersection points. Knowing that the equations represent a hyperbola and a circle also tells us, with no extra work at all, that there will be at most four solutions (x, y), since there will be at most four intersection points for a hyperbola and circle in the xy-plane.

George Polya advised his students that *if you can't solve the given problem, try solving an easier problem first.* Solving an easier problem may give you some ideas on how to solve the harder problem. You can make your problem easier in a variety of ways. You can change the size or scale of the problem. You can use fewer equations or variables, you can use smaller parameters, or you can change the constraints on the problem.

An important problem solving strategy that often works is "divide-and-conquer." This strategy involves breaking your problem into two or more separate and distinct cases. Then you solve each case separately. There are many ways to do this. You might consider even and odd numbers separately. Or you might consider prime numbers and composite numbers as distinct cases. For a geometry problem, you might break your problem into pieces by separately considering acute angles, obtuse angles, and right angles. Try dividing your problem into separate, distinct cases that cover all the possibilities. Examine each case separately and solve the problem for each case.

Look for patterns.

A successful outcome of the exploratory process is one or more useful observations, or patterns. Patterns are the keys to solving difficult mathematics problems. Keith Devlin even defined *mathematics* as "the science of patterns." If you see a pattern in your data, then you know you are on the right track. A solution to your problem is imminent.

What kind of patterns should you look for? When you are exploring a mathematics problem, such as plugging numbers into equations and tabulating data, certain kinds of patterns are especially relevant to mathematics. Look for the appearance of mathematically important quantities such as:

- Prime numbers (or composite numbers)
- Even (or odd) numbers
- Powers of 2 (or 3, etc.)
- Integers (or rationals, etc.)

- 0 or 1
- Mathematical constants like π, e, or γ

Does your data show any periodic behavior? Consider the following number sequence: 7, 23, 4, 2, 0, 6, 9, 13, 8, 0, 1, 1, 7, 19, 0, 7, 7, 7, 9, 0.... Do you see a pattern? One pattern is that every fifth number is zero. This observation could be the key to solving your problem. You should investigate whether every fifth number really is zero, and, if so, why?

Learn to recognize important number sequences such as the following. These number sequences show up frequently in mathematics problems. If you recognize one of these sequences, then you definitely have found an important pattern.

- Prime numbers: 2, 3, 5, 7, 11, 13, 17, 19, 23, ...
- Fibonacci numbers: 1, 1, 2, 3, 5, 8, 13, 21, 34, 55, ...
- Powers of 2: 1, 2, 4, 8, 16, 32, 64, 128, 256, 512, ...
- Perfect squares: 1, 4, 9, 16, 25, 36, 49, 64, 81, ...
- Factorials: 1, 2, 6, 24, 120, 720, 5040, 40320, ...
- Catalan numbers: 1, 1, 2, 5, 14, 42, ...

If your numerical investigations reveal a number sequence that you don't recognize, then look it up in *Sloane's Handbook of Integer Sequences*. An online version can be found at https://oeis.org/. *Sloane's Handbook* will tell you what mathematicians know about your number sequence.

An especially important pattern is *Pascal's triangle* (see *Problem 9* in Part II):

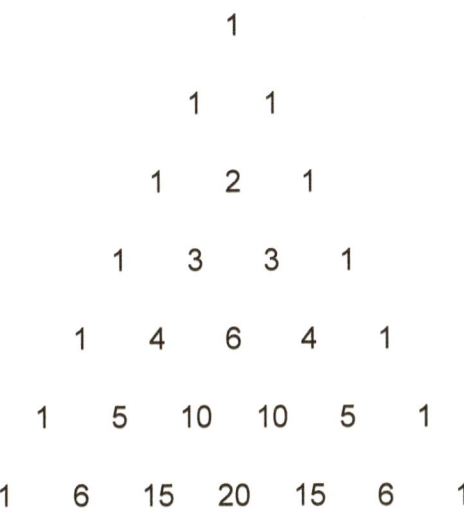

With the exception of the "1s," each number in Pascal's triangle is the sum of the two numbers directly above it. Note, for example, that $10 = 4 + 6$. If you are solving a combinatorics problem, you should *always* look for your numbers in Pascal's triangle. Why? We know from experience, and mathematical theory, that the answer to a combinatorics problem will always appear in Pascal's triangle if the answer is a positive integer. (Trivially, every positive integer is in Pascal's triangle. But *where and how* your numbers fall into Pascal's triangle is important.) Mathematicians know a great deal about the numbers in Pascal's triangle, and there are thousands of useful theorems and relationships that could help you solve your problem.

Watch out for false patterns.

Looking for and finding patterns is a central theme of our problem solving process. Most of the time, when we think we see a pattern it really is a pattern. That's why we should formulate patterns as conjectures and try to prove them. However, once in a while we think we see a pattern that isn't really there. These are *false patterns*. The human brain is so good at pattern recognition that it often sees patterns where none exist.

Here is a famous example of a false pattern. Consider the polynomial $f(n) = n^2 + n + 41$. This polynomial produces a prime number for $n = 1, 2, 3, ..., 39$. For example, $f(5) = 71$, which is prime. Based on the first 39 data points, we are tempted to form the conjecture that " $f(n)$ is a prime for all positive integers n." But this conjecture is false, because when $n = 41$ we get $f(41) = 41^2 + 41 + 41$, which is divisible by 41. So $f(41)$ is composite. ($f(40)$ is also composite.)

Go ahead and look for patterns. Most of the time, your patterns will be real. Just remember that you must always *prove* your observed patterns using sound mathematical arguments.

Chapter 5

Formulate Conjectures

When mathematicians do research and investigate problems, they look for patterns and make conjectures. A *conjecture* is an unproved mathematical statement that you *believe* to be true. Good conjectures precede their proofs. A conjecture is not a theorem. A theorem is a mathematical statement that has a proof. A conjecture is a statement that wants to become a theorem. If you *prove* your conjecture, then your conjecture will become a theorem.

When you examine your data, during your exploratory efforts, look for patterns. When you see a pattern, carefully describe your observed pattern. Write it down on paper. Write your observed pattern as a mathematical statement. This is your *conjecture*. It may help if you first write your conjecture in plain English, as if you wanted to explain it to your Yugoslavian grandma. Afterward, you can rewrite your conjecture using the technical language of mathematics to make it more precise. Let's look at a simple example.

What is the sum of the first n odd numbers? To get started, let's look at some small cases. Here is what we have:

$$1 = 1$$
$$1 + 3 = 4$$
$$1 + 3 + 5 = 9$$
$$1 + 3 + 5 + 7 = 16$$
$$1 + 3 + 5 + 7 + 9 = 25$$

Something interesting is going on here. The numbers on the right side of the equality sign appear to be perfect squares: $1 = 1^2$, $4 = 2^2$, $9 = 3^2$, $16 = 4^2$, and $25 = 5^2$. Is this always true? Does the pattern continue? I don't know. (I really do know, but let's pretend that I don't know). Let's state our observed pattern as a conjecture. In plain English, the conjecture is this:

Conjecture: The sum of the first n odd numbers is n^2.

Let's test the conjecture. Is the sum of the first 8 odd numbers equal to 8^2, or 64? It sure is: $1 + 3 + 5 + 7 + 9 + 11 + 13 + 15 = 64$. No amount of testing will suffice to prove a mathematical statement. However, a single failed test is a sufficient counter-example.

Now we are ready to rewrite our conjecture using the precise language of mathematics. The reason we want to do that is because it will actually make it easier to prove (or disprove) the conjecture. In fact, we will prove this conjecture in the next chapter. Then our conjecture will grow up and become a happy theorem.

The kth odd number can be written mathematically as $2k-1$. Try it out. The 5th odd number is $2(5)-1=10-1=9$. So, the sum of the first n odd numbers can be written as $\sum_{k=1}^{n}(2k-1)$. We can rewrite our conjecture more precisely as follows:

Conjecture: For positive integer n, $\displaystyle\sum_{k=1}^{n}(2k-1)=n^2$.

Now we have a precise mathematical statement that we can attempt to prove. And because we wrote the conjecture as a mathematical statement, using the language of mathematics, we can bring the full machinery of modern mathematics to bear on proving this statement.

Do you understand the mathematical statement? If not, use our old standby exploratory technique of *plugging numbers into the equation*:

$$\sum_{k=1}^{3}(2k-1)=(2\cdot1-1)+(2\cdot2-1)+(2\cdot3-1)=1+3+5=9=3^2.$$

After you explore a problem, identify patterns, and make conjectures, the next step is to *prove* your conjectures. We will address this topic in the next chapter.

Chapter 6

Prove Your Conjectures and Solve the Problem

After you explore your problem, identify patterns, and make conjectures about those patterns, the next step in the problem solving process is to prove your conjectures. If you can prove your conjectures, using mathematically sound arguments, then you can solve your problem.

The kinds of statements that we typically want to prove in mathematics are generally of a few flavors. First, there are statements of the form "If A, then B." We call these kinds of statement "implications." In geometry, for example, we have the famous Pythagorean theorem: *If a triangle has a right angle (90 degree angle), then the square of the length of the hypotenuse is equal to the sum of squares of the lengths of the other two sides.* A second kind of statement is an "existence" statement. Here, we want to prove that a mathematical object, having some set of interesting properties, actually exists. Then, if the object does exist, sometimes we want to prove that it is unique—that there is only one. This kind of proof is called a "uniqueness" proof.

How do we prove a mathematical conjecture? There are several different proof techniques that frequently appear in mathematics. We will briefly review the most important techniques. Our list of proof techniques, discussed here, is not exhaustive. Each special branch of mathematics has some specialized proof techniques. For example, in number theory we can sometimes prove that a given Diophantine equation (an equation whose solutions must be integers) has no positive integer solutions using a proof technique known at *Fermat's method of infinite descent.* Fermat's technique relies on an important property of positive integers that is good to know called the *Well Ordering Principle: Every non-empty set of positive integers contains a least (smallest) element.*

Here are the most important mathematical proof techniques. We will use some of these in Part II when we solve some problems.

Direct proof

A direct proof is pretty straightforward. To show, for example, that $a^2 - b^2 = (a+b)(a-b)$, all we have to do is multiply the expression on the right-hand side of the equality and show that it is the same as the expression on the left. The problem is that direct proofs aren't always available to us. Life is usually not that simple.

Proof by contradiction

Proof by contradiction is used quite frequently in modern mathematics. It has a fancy Latin name, *reductio ad absurdum*, which means "reduction to absurdity." You may have seen this method used to prove that $\sqrt{2}$ is irrational. Suppose we want to prove that some statement, A, implies another statement, B. Mathematicians write "A implies B" using the fancy notation $A \Rightarrow B$. This just means "If A is true, then B is true." To prove that A implies B using proof by contradiction, we assume that A is true and B is false. Next, using a clever argument, we deduce a contradiction of some sort. Thus, if A is true, B cannot be false. Statement B must be true.

Let's prove, by contradiction, that there is no even prime number greater than 2. Assume, to the contrary, that p is an even prime number greater than 2. Because p is even, it has the form $p = 2n$, and, since p is greater than 2, n is a positive integer greater than 1. But then p is divisible by both 2 and $n > 1$. This is a contradiction. A number greater than 2 having two or more non-unit factors is composite. This contradicts our assumption that p is prime. Therefore, there is no even prime number greater than 2.

Proof by contrapositive

It is a fact of logic that the statement "*A* implies *B*" is logically equivalent to the statement "Not *B* implies not *A*." Let's take a simple everyday example. The statement "If it rains on me, then I get wet" is logically the same as the statement "If I do not get wet, then it does not rain on me." Sometimes it is easier to think about a mathematical statement if we rephrase it as the equivalent contrapositive statement. Using this method, we prove that "*A* implies *B*" by showing that "If *B* is false, then *A* is false."

Suppose we have a triangle with sides *a*, *b*, *c*, and $a \leq b \leq c$. Can we prove that if c^2 is not equal to $a^2 + b^2$ then the triangle is not a right-angled triangle? Of course we can. The contrapositive statement is just the Pythagorean theorem, which we already know is true: *If the triangle is right-angled, then $c^2 = a^2 + b^2$, where c is the hypotenuse (side opposite the right angle).*

Working backwards

To prove that "*A* implies *B*," assume that *A* is true and that *B* is true. Work backwards from *B* to *A*. If the backward mathematical steps are uniquely reversible, then you can follow them in the reverse direction from *A* to *B*.

Proof by construction

If the mathematical statement B has the words "there exists," or "there is," then to prove that "A implies B," first assume that A is true. Then construct an object of type B based on the assumption that A is true.

Let's show, by construction, that if p and q are any two rational numbers (ratios of whole numbers), with $p < q$, there exists another rational number, r, between p and q. The proof is by construction. One such number is $r = (p+q)/2$.

Proof of uniqueness

If B is a statement of uniqueness ("there exists a unique…"), then to prove that "A implies B," first assume that A is true. Then assume that there are two different objects of type B, and show that this assumption results in a contradiction.

Nonzero real numbers have multiplicative inverses. For example, the multiplicative inverse of $x = 5$ is $y = 1/5$, since $xy = 1$ and 1 is the multiplicative identity. Let's prove that a multiplicative inverse for a real number, if it exists, is unique. Suppose a real number x has an inverse y. Then, by the definition of multiplicative inverse, $xy = yx = 1$. Now suppose that x has a second inverse z. Then $xz = zx = 1$. Using the associative law of multiplication, we have $z = 1 \cdot z = (yx)z = yxz = y(xz) = y \cdot 1 = y$. Therefore, $z = y$. So, the multiplicative inverse, if it exists, is unique.

Proof by counterexample

Some mathematical statements can be proved (or disproved) simply by producing a counterexample. Here is a simple example. Suppose you had the conjecture that all prime numbers are odd. Looking at the primes 3, 5, 7, 11, 13, 17, 19, …, it certainly looks like all prime numbers are odd. However, the conjecture is false. All we need to do is produce a single counterexample. The number 2 is an even prime number. In fact, 2 is the only even prime number.

Our final proof technique is used quite often in mathematics, especially in number theory and combinatorics. It is so important that we will give an example by proving our conjecture from Chapter 5. The other proof techniques will show up in Part II.

Mathematical induction

Mathematical induction looks strange the first time you see it. Proof by mathematical induction is like tipping over a line of dominoes. Imagine that you have some dominoes all lined up in a row. If the nth domino falls over and hits the next domino, then the $(n + 1)$st domino will also fall over. However, to get the process started, you must first knock over a domino. And if you want all of the dominoes to fall over, then you must knock over the first domino. Then the second will fall, and then the third, and so on, until all the dominoes fall over.

Mathematical induction is best learned by example. So let's prove our conjecture from Chapter 5 using this method of proof. I'm going to state the conjecture as a "Theorem" because we will successfully prove that it is a true statement.

Theorem: *For positive integer* n, $\displaystyle\sum_{k=1}^{n}(2k-1)=n^2$.

Proof by mathematical induction: Let $\displaystyle f(n)=\sum_{k=1}^{n}(2k-1)$. We want to show that $f(n)=n^2$. First, like knocking over the first domino, we must show that $f(1)=1^2$. We have

$$f(1)=\sum_{k=1}^{1}(2k-1)=(2\cdot1-1)=2-1=1=1^2.$$ So the first step,

which we call the "induction basis," is true. Now comes the clever part. We want to show that if we knock over domino number n then it will knock over domino number $n+1$. This means that we want to show that *if* $f(n)$ *is true,* *then* $f(n+1)$ *is* also true. This means that we can write out the equation for

$f(n+1)$, *assume* that $f(n) = n^2$, and see if we get $f(n+1) = (n+1)^2$. Here we go:

$$f(n+1) = \sum_{k=1}^{n+1}(2k-1)$$
$$= \sum_{k=1}^{n}(2k-1) + (2(n+1)-1)$$
$$= \sum_{k=1}^{n}(2k-1) + 2n+1$$
$$= n^2 + 2n+1$$
$$= (n+1)(n+1)$$
$$= (n+1)^2. \quad \square$$

Notice that we used the assumption that $\sum_{k=1}^{n}(2k-1) = n^2$ to prove that $\sum_{k=1}^{n+1}(2k-1) = (n+1)^2$.

Now, since $f(1) = 1^2$, and since the truth of $f(n)$ implies the truth of $f(n+1)$, it follows that $f(2) = 2^2$. But since $f(2) = 2^2$, this implies that $f(3) = 3^2$, and so on, until all the dominoes fall over.

Mathematical tactics

After you identify significant patterns, formulate conjectures, and prove them, you will need to solve the final problem. This involves putting all the puzzle pieces together. Often, to be successful, you will need to use special techniques or tricks—mathematical *tactics*—to solve the problem. Examples of tactics are things like factoring polynomials, multiplying by complex conjugates, or examining even and odd cases separately. There are far too many tactics to discuss all of them here. The best way to learn tactics is to solve lots of mathematics problems. We will see many practical applications of tactics when we solve mathematics problems in Part II. Appendix C provides a reference list of many commonly used mathematical tactics. You can even use Appendix C as a kind of checklist each time you solve a mathematics problem. The checklist might suggest a mathematical tactic that you forgot to try.

Chapter 7

Verify Your Solution

A good discipline to follow is to always verify your solution. In some cases, you may have made a mistake. In other cases, you made no mathematical mistake at all, but you may have extraneous solutions that don't satisfy the problem statement.

Technically, what I call "verify your solution" has two parts: *Verify your answer*, and *verify your solution*. If the problem asks for an answer, like a number, you should check to see that the number actually works—that it solves the problem. In addition, since mathematicians want to see well-reasoned solutions to a problem, it is more important that your *solution* is correct. Verifying the correctness of your solution means that you should verify your reasoning process, your arguments, and your logic.

How can you verify your solution? There are several methods. The strongest method is to show that your logic and reasoning is impeccable. You will need to carefully retrace your reasoning steps and look at all the small details. It helps if other mathematicians can peer-review your work and certify its correctness.

A second approach, which is more practical, is *numerical testing*. No amount of numerical testing can prove a theorem, but a single counterexample will negate a hypothesized statement. Numerical testing—plugging numbers into your equations or formulas—can also go a long way toward helping you understand what a theorem means.

A third verification approach is to look at the *internal consistency* of your solution or theorem. This is the "fit test." Does it fit within the existing framework of modern mathematics? Does your result contradict another well-established mathematical theorem, or does it fit well and harmonize with other known results? If you prove a theorem that contradicts the Fundamental Theorem of Arithmetic (uniqueness of prime factorization), then your theorem is wrong. It must be wrong, even if you don't know why, because the Fundamental Theorem of Arithmetic has been proved many times, by many mathematicians, in many different ways. Your result, if true, must be internally consistent with existing mathematical theory.

Let's look at a problem that shows us why we must always verify our solution.

Find the distinct solutions of the equation $\left| x - \left| 2x + 1 \right| \right| = 4$.

Here, the absolute value function is defined as follows:

$$|x| = \begin{cases} x, \text{ for } x \geq 0, \\ -x, \text{ for } x < 0. \end{cases}$$

We can use this definition to simplify $\left| x - \left| 2x + 1 \right| \right| = 4$.

The equation $\left|x-|2x+1|\right|=4$ can be written as (1) $x-|2x+1|=4$

or (2) $x-|2x+1|=-4$. There are two possibilities that follow

from the definition of the absolute value function. Equation (1)

can be written as $|2x+1|=x-4$, which then implies two

possibilities: (3) $2x+1=x-4$ or (4) $2x+1=-(x-4)$. Solving

equation (3) for x gives $x=-5$. Solving equation (4) for x gives

$x=1$. Next, examining equation (2), we see that equation (2)

can be written as either (5) $2x+1=x+4$ or (6) $2x+1=-(x+4)$.

Solving equation (5) for x gives $x=3$. Solving equation (6) for x

gives $x=-5/3$. In summary, we have found what looks like

four distinct solutions: $x=-5, 1, 3, -5/3$. Is this correct?

Although our mathematical analysis was correct, only $x=3$ and

$x=-5/3$ are valid solutions of the original equation

$\left|x-|2x+1|\right|=4$. The other two solutions, $x=-5$ and $x=1$ are

extraneous solutions. We can see this by plugging them back

into the original equation:

$$\left|-5-|2(-5)+1|\right|=\left|-5-|-9|\right|=|-5-9|=|-14|=14,$$
$$\left|1-|2(1)+1|\right|=\left|1-|3|\right|=|1-3|=|-2|=2.$$

The moral to the story is to *always check your answers and verify*
your solutions.

Chapter 8

Polish the Stone

Showmanship is important. After you solve a mathematics problem, you should "polish the stone." Clean up your solution, eliminate unnecessary distractions, focus on the essential aspects, and write it down. Write your solution as if you were Renoir placing colorful brush strokes on canvas. State your theorems and write your proofs in the most elegant way possible. Be an artist. Create mathematical beauty.

People who think of mathematics as little more than a sophisticated form of accounting are often surprised by the idea of mathematics as a fine art. You might think, wrongly of course, that a mathematician has no creative latitude for placing the "brush strokes" on the "canvas." Mathematics becomes a fine art when we learn to play with it like an artist. Let's look at an example of how you might play with mathematics like an artist and polish the stone. There is no right or wrong here. Just have fun and play with the mathematics.

Suppose you prove that a certain function $f(n)$, defined on the positive integers, has the form $f(n) = \dfrac{1}{2n+1}$. You could stop here. But why stop here? What does this function mean? One interpretation is that $2n+1$ is an odd number. What else can we do? Can we do something fun and creative? Here is one idea. If you realize that $1+2+3+...+n=n(n+1)/2$ and that $1^2+2^2+3^2+...+n^2=n(n+1)(2n+1)/6$, then you can rewrite $f(n)$ as a statement about ratios of sums of positive integers:

$$3 \cdot f(n) = \frac{1+2+3+...+n}{1^2+2^2+3^2+...+n^2}.$$

The great 18th century mathematician Leonhard Euler often did this kind of creative manipulation, and he was very good at it. *Beauty is worthy of contemplation.* Beautiful theorems also tend to be important theorems. The driving force in your study of mathematics as a fine art should be the pursuit of mathematical beauty.

When we "polish the stone," we present our mathematical reasoning and results in a "clean," refined, and elegant manner. We formulate our results so that they express the greatest amount of information with the least amount of effort. This is the principle of *economy of force*. Mathematics becomes powerful when simple statements encompass much territory.

Ordinarily, it is standard practice for mathematicians to avoid mentioning their exploratory process. This is why mathematicians sometimes appear to be geniuses. They state a problem, propose a clever idea, and solve the problem. It has all the appearances of a magician pulling a rabbit out of a hat. It only seems that way, because you do not see behind the magician's curtain. A mathematician may spend many weeks, filling waste paper baskets with failed attempts, before finally stumbling on a brilliant solution. Then the mathematician "polishes the stone." She eliminates waste and fluff and presents a mathematical argument that is as polished and professional as a *Cirque du Soleil* performance. You should do the same.

Always document your solution in writing. The process of writing will help clarify your thoughts and identify mistakes. Writing forces you to think through all the small details and explain them to someone else. If you can't explain your mathematics in written form, then you don't really understand your own work. You should consider writing—mathematical writing—to be the final and ultimate step in your problem solving process.

Chapter 9

Reflect and Learn

If you want to become a better mathematics problem solver, you should reflect and learn from each problem you tackle. After you solve a problem, carefully analyze your solution. Did you have any difficulties or mental blocks? What did you do right? What could you have done better? Look for key ideas, concepts, principles, and methods that allowed you to solve the problem. Document what you have learned. Those things will be useful when you solve future problems. Also, look for any theorems that might help you solve the problem more efficiently.

Besides reflecting on your solution, see if there is another solution. Problem books usually have solutions to problems in the back of the book. Study other people's solutions to see if you can learn something from them. Usually, there is more than one valid way to solve a mathematics problem. Other people may have found better ways to solve your problem. By studying their solutions, you can learn and improve your ability to solve mathematics problems.

A good practice is to collect good problems. In the course of working problems, some of them will be particularly interesting, illuminating, or instructive to you. Some may even be beautiful.

You should save, collect, and study these problems. I like to keep good problems on 5 x 8 index cards. Each card has three sections: the *problem*, *key ideas*, and the *solution*. As your collection grows, you can begin to classify the problems by type: number theory, combinatorics, algebra, logic, and so on. Here is a sample problem card from my collection:

No. 27, Congruences	**Number Theory**

Problem: Can the number 19^{19} be represented as the sum of a positive integer cubed and a positive integer to the fourth power? $19^{19} = m^3 + n^4$ for $m, n \in Z^+$?

Key Ideas: Consider integers modulo 13.

Solution: No. It cannot be done. When we consider integers modulo 13, we find that $m^3 \equiv 0, 1, 5, 8,$ or 12 modulo 13. Likewise, $n^4 \equiv 0, 1, 3,$ or 9 modulo 13. Therefore $m^3 + n^2$ can be congruent to 0, 1, 2, 3, 4, 5, 6, 8, or 9, but not 7 modulo 13. However, $19 \equiv 6$ modulo 13; so $19^{19} \equiv 6^{19} \equiv 7$ modulo 13. So it is not possible. Note, by Fermat's Theorem, that $6^{12} \equiv 1$ modulo 13. So $6^{19} \equiv 6^7 \equiv 7$ modulo 13 (after further simplification using dirty tricks).

Part II

SOLVING MATHEMATICS PROBLEMS
BY THE FIREPLACE

Part II Introduction

We will now work through and really try to understand some good, challenging mathematics problems. Most problem solving books present a problem along with some clever, ingenious solution. Even if you understand the solution, you invariably feel stupid, because you feel like there is no chance that you could be so clever. Most of that feeling is just an illusion. The person who solved the problem probably did not begin with a simple, ingenious idea. Instead, he had to experiment, struggle, and suffer, perhaps for a long time, before finally stumbling upon an elegant solution. Generally, in mathematical problem solving, ugly solutions precede beautiful solutions. A solution is made to appear beautiful only after you have done a lot of hard work to discover the ugly solution. Then, you "polish the stone," clean up your written solution, and make the solution look simple and elegant. So instead of simply presenting a hard problem followed by a divinely inspired solution, let's sit around the fireplace like you are having a friendly chat with your dear uncle and do some mathematical exploration. As we explore these problems, look for key ideas and insights that reveal the final solution.

Problem 1

Sum of Roots

Let $f(x)$ be a real-valued function defined on the real numbers. If $f(x)$ has five distinct real roots and $f(5-x) = f(5+x)$, find the sum of the roots.

This problem looks intimidating. We are asked to find the sum of the roots, but we don't even know what the function $f(x)$ looks like. It seems like an impossible problem.

It would be really nice if the function $f(x)$ were a polynomial, for then we could use the fact that there are five distinct roots, along with the Fundamental Theorem of Algebra, to build a general fifth degree polynomial. That would at least give us some place to start. But the problem does not specify that $f(x)$ is a polynomial. We are only told that the function is a real-valued function with five distinct roots. The function might look like anything imaginable. Wow! We feel lost. Where do we start?

Do we really understand the problem? First, the problem does not ask us to actually find the five roots. Instead, we are asked to find the *sum* of the five roots. So, maybe we can treat the sum of the roots as a discrete entity, without having to actually determine the individual roots x_1, x_2, x_3, x_4, and x_5. Second, since the function $f(x)$ is defined on the real numbers, we don't need to worry about imaginary or complex numbers. Third, since we are only given the relationship $f(5-x) = f(5+x)$, the best strategy for solving this problem might be to focus on the *relationship* rather than attempting to determine what function $f(x)$ might actually be.

What is our brute-force-dumb solution? Unfortunately, for this problem, I'm at a loss of ideas. Since we don't know $f(x)$, we can't even plug numbers into the function to look for patterns in the data. We could try some curve sketching, and I wasted a lot of time doing that, but it doesn't help much.

Let's explore the problem by first focusing on the given statement $f(5-x) = f(5+x)$. It would be nice to know, and possibly useful, the relationships involving simple, basic things like $f(0)$, $f(x)$, and $f(-x)$. Another idea is to see if $f(x)$ is a *periodic function* by checking if $f(x) = f(x+T)$ for some number T, called the *period*. I would rule out this possibility, though, because we are told that $f(x)$ has five distinct roots. If $f(x)$ were periodic, then the existence of one root would imply infinitely many others, since $f(x_1) = f(x_1 + T) = 0$, and then $f(x_1 + T) = f((x_1 + T) + T) = f(x_1 + 2T) = 0$, and so on.

Let's see if we can find a relationship for $f(x)$. I don't like the arguments $5-x$ and $5+x$. So, let's make the substitution $u = x+5$, or $x = u-5$. Substituting this into the given relationship, we get $f(u) = f(10-u)$, or simply $f(x) = f(10-x)$ if we want to keep writing everything in terms of the variable x. This last step is just a relabeling of u by x. Good problem solving involves a lot of "gear changing." The reason why I like this relationship better than the given one is because we now have $f(x)$ by itself on the left-hand side of the new equation. Now, let's think for a moment. We know that $f(x)$ has five distinct roots, say x_1, x_2, x_3, x_4, and x_5. So plug those roots into the equation $f(x) = f(10-x)$ and see what happens:

$$\begin{aligned} f(x_1) &= f(10-x_1) = 0, \\ f(x_2) &= f(10-x_2) = 0, \\ f(x_3) &= f(10-x_3) = 0, \\ f(x_4) &= f(10-x_4) = 0, \\ f(x_5) &= f(10-x_5) = 0. \end{aligned}$$

These equations tell us that not only are the five distinct roots x_1, x_2, x_3, x_4, and x_5, but the five distinct roots are also, in some order, $10-x_1$, $10-x_2$, $10-x_3$, $10-x_4$, and $10-x_5$. So, the sum of the roots must be:

$$x_1 + x_2 + x_3 + x_4 + x_5 = (10 - x_1) + (10 - x_2) + (10 - x_3) + (10 - x_4) + (10 - x_5)$$
$$= 50 - (x_1 + x_2 + x_3 + x_4 + x_5)$$

If we let $S = x_1 + x_2 + x_3 + x_4 + x_5$, then $S = 50 - S$, or $2S = 50$.

Therefore, the final answer is $S = 25$. The sum of the five distinct real roots of $f(x)$ is 25. $\square$

Problem 2

Product of Tangents

Find the value of $P = \tan(15) \cdot \tan(30) \cdot \tan(45) \cdot \tan(60) \cdot \tan(75)$, where the arguments are given in degrees, not radians.

This problem is a great example of the distinction that mathematicians make between an "answer" and a "solution." Finding an answer is easy. We simply grab a pocket calculator and calculate the numerical answer. If you did that on a college mathematics test, the professor would give you an "F." Why? Because what we really want is a "solution" to the problem. We want to see an insightful, logical, and beautiful way to solve the problem without resorting to a calculator or table of trigonometry functions. Anybody can do that. It takes an artist to find a beautiful solution.

How do we begin? Let's start with the brute-force-dumb solution. Grab a calculator, plug in the numbers, hit the tangent key, and calculate the product. Surprisingly, the answer we get is "1." Wow! That's no coincidence. The math god must have arranged it that way on purpose. When you get a really nice answer, like 1 or 0, you must believe that something really interesting is going on with your problem. It is often easier to find the solution to a problem if you already know the answer. Knowing that the answer is $P = 1$ points us in the right direction.

Since the answer to the problem is 1, which is a really nice answer, we should ask ourselves how it could be 1. Tangents, and other trigonometric functions like sine and cosine, give us ratios of sides in right-angled triangles. So, the tangent of an angle will be a ratio, like $\frac{a}{b}$. Since we are multiplying a bunch of tangents, and since we already know that the answer is 1, these ratios must somehow cancel each other out. A ratio like $\frac{a}{b}$ must be multiplied by $\frac{b}{a}$ so that $\left(\frac{a}{b}\right)\left(\frac{b}{a}\right) = 1$. This observation suggests that a solution to the problem must involve *pairing* up the tangent functions in some clever way. But how?

In a right-angled triangle, one angle is 90 degrees. The other two angles are, say, θ and φ. Since the sum of the angles in any triangle is 180 degrees, we must have $\theta + \varphi = 90$, or simply $\varphi = 90 - \theta$. Then, if the tangent of θ is $\frac{a}{b}$, the tangent of φ, which is the other angle, must be $\frac{b}{a}$. Draw a picture:

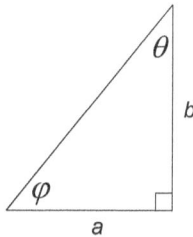

Since $\left(\dfrac{a}{b}\right)\left(\dfrac{b}{a}\right) = 1$, we have $\tan(\theta) \cdot \tan(\varphi) = \tan(\theta) \cdot \tan(90 - \theta)$

$= 1$. That's exactly what we need. Now we know how to proceed. We pair up the tangent functions that have complementary angles (angles that sum to 90 degrees) as follows:

$$(\tan(15) \cdot \tan(75)) \cdot (\tan(30) \cdot \tan(60)) \cdot \tan(45).$$

Well, we can pair up 15 degrees with 75 degrees, since $15 + 75 = 90$. Likewise, we can pair up 30 and 60, since $30 + 60 = 90$. What about 45? We got lucky here. The tangent of 45 degrees is just 1. So, using the *pairing* trick, along with the fact that $\tan(\theta) \cdot \tan(90 - \theta) = 1$ and $\tan(45) = 1$, we get the solution:

$(\tan(15) \cdot \tan(75)) \cdot (\tan(30) \cdot \tan(60)) \cdot \tan(45) = 1.$ □

Note that, with the exception of the tangent of 45 degrees, we didn't need to know the actual values of the tangents. No calculator or trigonometry tables were necessary. When you write up your solution for publication, you don't tell people about your sneaky investigative techniques. You state the problem, declare that you should use a clever pairing technique, and solve the problem. People will think you are a genius.

Problem 3

Monic Polynomial of Degree 4

A monic polynomial $f(x)$ of degree four has the following values: $f(1) = 10$, $f(2) = 20$, and $f(3) = 30$. Find the value of $f(12) + f(-8) - 19000$.

How can we solve this problem? Our first thought is to find the polynomial $f(x)$. That seems reasonable, since knowing $f(x)$ would allow us to easily calculate $f(12) + f(-8) - 19000$. A *monic polynomial*, by definition, is a polynomial whose leading coefficient is one. So the general form of our fourth degree monic polynomial is $f(x) = x^4 + ax^3 + bx^2 + cx + d$ for some undetermined real (or maybe complex?) coefficients a, b, c, and d. There is a theorem that may be useful here.

Theorem 3.1. *A polynomial $f(x)$ of degree n is uniquely determined by $n+1$ different x-values.*

We have a potential problem. Our polynomial has degree four, but we only know *three* of its values, not five! Of course, it's not quite that bad. Since the polynomial is monic, we really only have four unknowns, a, b, c, and d, but we still have the problem that we know only three values of the function. What do we do? Let's just have faith that things will somehow work out and move forward. Having blind faith, throwing caution to the wind, and advancing on the enemy is not logical, but it's a perfectly valid strategy in mathematics. Here is what we have so far after using the given information $f(1) = 10$, $f(2) = 20$, and $f(3) = 30$:

$$f(1) = a+b+c+d+1 = 10$$
$$f(2) = 8a+4b+2c+d+16 = 20$$
$$f(3) = 27a+9b+3c+d+81 = 30$$

Rearranging the equations, we obtain a system of three simultaneous linear equations in four unknowns:

$$a+b+c+d = 9$$
$$8a+4b+2c+d = 4$$
$$27a+9b+3c+d = -51$$

Since we have fewer equations than unknowns, and this seems like a uniquely unsolvable condition, let's just pretend like we know d. Just use some wishful thinking and hope it all works out. Pretending like d is just another number, we rewrite the system of equations so that it looks like we have three equations in three unknowns (a, b, and c):

$$a+b+c = 9-d$$
$$8a + 4b + 2c = 4 - d$$
$$27a + 9b + 3c = -51 - d$$

If you like matrices, you can also write these equations in matrix form as follows:

$$\begin{bmatrix} 1 & 1 & 1 \\ 8 & 4 & 2 \\ 27 & 9 & 3 \end{bmatrix} \begin{bmatrix} a \\ b \\ c \end{bmatrix} = \begin{bmatrix} 9-d \\ 4-d \\ -51-d \end{bmatrix}$$

At this point, we have a standard exercise in linear algebra. There are more elegant methods, but a straightforward method is to solve the first equation for a in terms of b and c. Substitute that for a in the second equation, and solve the second equation for b. Then substitute for a and b in the third equation and solve for c. That will give you c. Then you can work backwards to get b and a. We'll skip the details, but you can actually solve the system for the unknowns to get $a = \dfrac{-d}{6} - 6$, $b = d + 11$, and $c = 4 - \dfrac{11d}{6}$. Now what? Well, just substitute these values for a, b, and c into the polynomial $f(x) = x^4 + ax^3 + bx^2 + cx + d$ to get $f(x) = x^4 + (-d/6 - 6)x^3 + (d+11)x^2 + (4 - 11d/6)x + d$. It looks like a complicated expression, but watch what happens when we evaluate $f(12) + f(-8) - 19000$. We get

$$f(12) + f(-8) - 19000 = 12000 - 165d + 165d + 7840 - 19000$$
$$= 840. \ \square$$

The unknown d magically cancelled out, and we have our first mathematical proof that a benevolent God exists.

Problem 4

No Negative Roots

Prove that the polynomial equation $x^4 - 5x^3 - 4x^2 - 7x + 4 = 0$ has no negative real roots.

    This is one of my favorite problems. It's a good teaching problem because it involves several important problem solving ideas and techniques. We may remember that the *roots* of a polynomial $f(x)$ are the values of x that make $f(x) = 0$. The Fundamental Theorem Algebra, which is the most important theorem in algebra, tells us that a polynomial of degree n has precisely n roots in the domain of complex numbers (allowing for the possibility of repeated roots). A brute-force-dumb approach to solving this problem would be to simply find the roots of this polynomial. In fact, if I use my TI-89 calculator, it tells me that the four roots are $x = 0.4211852391$, 5.865581749, $-0.643383494 + 1.097800278\ i$, and $-0.643383494 - 1.097800278\ i$. The last two roots are complex, since they involve the imaginary number $i = \sqrt{-1}$. Only the first two roots are real, and they are clearly not negative numbers. That's a brute-force-dumb solution to the problem. But it does tell us the given statement is true. There are no negative real roots for the polynomial.

How can we solve this problem in a more elegant fashion? First, note that this is a "to prove" kind of problem. We are not actually being asked to find the roots of the polynomial. That's a red herring. We are asked to *prove* that the polynomial has no negative real roots. How can we prove that?

A good way to prove problems like this is to use *reductio ad absurdum* (Latin for "reduction to absurdity"). This is just the proof technique of *proof by contradiction*, and mathematicians use this method quite often. The way it works is that we *assume* that the given polynomial does have a negative root, and then we demonstrate that this assumption leads to a contradiction. Since we will get a contradiction, our assumption must have been false. Therefore, the equation has no negative real roots.

Suppose that x were a negative root. What would happen? We would substitute x into the polynomial, and we would have to calculate the various powers of x, like x^4, x^3, x^2, and x^1. So, what happens when we calculate the powers of a negative number? Let's look at a typical case, say -1. The powers of -1 are $(-1)^4 = +1$, $(-1)^3 = -1$, $(-1)^2 = +1$, and $(-1)^1 = -1$. This is an interesting pattern. We see that $(-1)^{even} = +1$ and $(-1)^{odd} = -1$. In fact, if we think about it, this observation is true for the powers of any negative number because "*negative* × *negative = positive*" and "*negative × negative × negative =*

negative." This suggests that we should focus on the *parity* (evenness or oddness) of the powers of x.

Now we will use the principle of "divide-and-conquer." We will split the equation based on the parity of the powers of x. Let's get the odd powers of x on one side of the equality, and put the even powers on the other side:

$$5x^3 + 7x = x^4 - 4x^2 + 4$$

We have odd powers of x on the left-hand side and even powers of x on the right-hand side. Hopefully, when x is negative, we will get some kind of contradiction. Let's do one more thing that may help us here. Let's factor the right-hand side of the equation. Why? This is intuition at work. *Always try to factor your polynomials.* Just do that. Make it a habit. If you see a polynomial, factor it. If you see a donut, eat it. It's as simple as that. We then obtain the following expression:

$$5x^3 + 7x = (x^2 - 2)^2$$

Suppose that x is negative. Then the left-hand side will be negative, because odd powers of a negative number are always negative. What about the right-hand side? On the right-hand side we have a square, and *no square is negative*. So, the right-hand side will be positive. Thus, we have a contradiction: *negative = positive*. Therefore, our original assumption that x is a negative root is false! The given polynomial can have no negative real roots. Our proof is done. Let's eat a donut. □

Problem 5

A Periodic Function

Prove that if $f(x)$ is a non-constant real-valued function such that for all real x, $f(x+1)+f(x-1)=\sqrt{3}f(x)$, then $f(x)$ is periodic.

This kind of problem is a problem involving *functional equations*. We are given a relationship for some function, $f(x)$, but we don't what the function is. The problem asks us to prove that the unknown function $f(x)$ is periodic.

Let's go back to basics. What is a periodic function? A function $f(x)$ is said to be *periodic*, with period T, if there exists some constant T such that $f(x+T)=f(x)$. Our task is now clear. We must show that $f(x)=f(x+T)$ for some real number T.

Since the function $f(x)$ is unknown, and there may not be much hope of even determining what it is, we must instead focus on the given *relationship*. This is typical with functional equations. Use the relationship that is given to you, and try using some combination of substitutions and back-substitutions to solve the problem. Note that we really don't need to know what the function $f(x)$ is anyway. We just need to show that

$f(x) = f(x+T)$ for some real number T. That may actually be an easier problem. There is no guarantee that the unknown period T is a positive integer. It may not be. But let's just assume that T is a positive integer and see what happens.

We can rewrite the given equation as follows:

$$f(x+1) = \sqrt{3}f(x) - f(x-1).$$

That gives us $f(x+1)$, which unfortunately is not equal to $f(x)$.

What about $f(x+2)$? Surprisingly, we can find $f(x+2)$ because we can replace x by $x+1$ in $f(x+1)$ to get

$$\begin{aligned} f(x+2) &= f((x+1)+1) \\ &= \sqrt{3}f(x+1) - f((x+1)-1) \\ &= \sqrt{3}f(x+1) - f(x). \end{aligned}$$

We see that $f(x+2)$ is not equal to $f(x)$. That's too bad. So, we just keep on calculating and hope that our perseverance pays off. Let's calculate $f(x+3)$. Since we know $f(x+1)$ and we now know $f(x+2)$, we can calculate $f(x+3)$:

$$f(x+3) = f((x+1)+2)$$
$$= \sqrt{3}f(x+2) - f(x+1)$$
$$= \sqrt{3}(\sqrt{3}f(x+1) - f(x)) - f(x+1)$$
$$= 3f(x+1) - \sqrt{3}f(x) - f(x+1)$$
$$= 2f(x+1) - \sqrt{3}f(x).$$

We still don't have $f(x) = f(x+T)$. Either this problem takes persistence or we are on the wrong track. Let's keep going, using back-substitution as necessary, and calculate $f(x+4)$, $f(x+5)$, and $f(x+6)$. Eventually, after a lot of hard work, we find that $f(x+6) = -f(x)$. This is almost what we want, except that it has a minus sign in front of $f(x)$. Are we stuck? Let's not give up yet. Maybe we can use this last discovery.

We know that $f(x+6) = -f(x)$, so if we replace x by $x+6$, we get the following:

$$f(x+12) = f((x+6)+6) = -f(x+6) = -(-f(x)) = f(x).$$

At last we are done. We have shown that $f(x+12) = f(x)$, so the unknown function $f(x)$ is periodic with period 12. □

Sometimes the solution to your problem requires little more than a good idea and a lot of hard work. When I solved this problem, I was almost ready to give up by the time I calculated

$f(x+5)$. The winning calculation occurred when I got to $f(x+6)$. Don't give up easily. Be persistent!

Problem 6

Powers of Three

Find the 50th smallest positive integer that can be written as the sum of distinct powers of 3 with nonnegative integer exponents.

A good place to start with this problem is to see if you can find a few of the required numbers. What do they look like? One example of such a number is $3^0 + 3^2 + 3^5 = 253$. The number 253 has the desired properties, since it is the sum of distinct powers of 3 with nonnegative integer exponents. It's OK to have an exponent of zero, because zero is nonnegative. Can we find another such number? Here is another one: $3^1 + 3^2 + 3^3 + 3^7 = 2226$. By looking at a few numbers that have the desired properties, we are hoping to see some kind of pattern, or insight, which will help us solve the problem.

A key observation is that the representations $3^0 + 3^2 + 3^5$ and $3^1 + 3^2 + 3^3 + 3^7$ look suspiciously like *base 3* representations of numbers. A positive integer n is represented in base 3 by the following form:

$$n = a_k 3^k + a_{k-1} 3^{k-1} + \ldots + a_2 3^2 + a_1 3^1 + a_0 3^0,$$

where the coefficients a_j are either 0, 1, or 2. In our problem, however, the coefficients are either 0 or 1, not 2. Our example number 253 looks like this:

$$1 \cdot 3^5 + 0 \cdot 3^4 + 0 \cdot 3^3 + 1 \cdot 3^2 + 0 \cdot 3^1 + 1 \cdot 3^0 = (1, 0, 0, 1, 0, 1)_3.$$

So, a brute-force-dumb solution to the problem would be to simply list all the numbers of the required form in order from smallest to largest, and then find the 50th smallest one in the list. But when we do this, all of the base 3 representations *look like* binary numbers! For example 253, shown above, looks like $(1, 0, 0, 1, 0, 1)_3$ in base 3. And all those 1's and 0's remind us of binary numbers. This observation suggests a clever idea. Let's just list all the binary numbers, in order, from 1 to 50:

$$1 = 1$$
$$2 = 10$$
$$3 = 11$$
$$4 = 100$$
$$5 = 101$$
$$...$$
$$50 = 110010$$

Now, pretend like 110010 (binary for the number 50) is actually a *base 3* number. Then

$$(110010)_3 = 1 \cdot 3^5 + 1 \cdot 3^4 + 0 \cdot 3^3 + 0 \cdot 3^2 + 1 \cdot 3^1 + 0 \cdot 3^0$$
$$= 3^5 + 3^4 + 3^1$$
$$= 327.$$

The answer to the problem is 327. □

Let's reflect on how we solved this problem. At first, we had no idea how to begin. So we started by finding a couple numbers of the required type. These were numbers like 253 and 2226 that can be represented as the sum of distinct powers of 3 with nonnegative exponents. Indeed, we "built" the numbers so that they would have that property. Then we noticed that the representations as distinct powers of 3 looked like base 3 representations, except that the coefficients are 0 and 1, never 2. When we represented these "base 3" numbers as a vector of coefficients, such as $(1, 0, 0, 1, 0, 1)_3$, we *shifted gears—changing our mental perspective*—and pretended like the numbers are actually *binary* numbers. At that point, the solution became apparent: Simply list all the binary numbers, in order, from 1 to 50. Pick the 50th number on the list, and pretend like it is a base 3 representation. Finally, convert the base 3 representation to decimal, and you get the answer, 327.

Problem 7

Integers in Sequence

An urn contains n balls numbered 1, 2, ..., n. You randomly remove balls from the urn, one at a time, until the urn is empty. What is the probability that throughout this process the numbers on the balls which have been removed is a sequence of consecutive integers?

First, we need to understand the problem. What does a "successful" process look like? Let's construct one. Suppose we reach into the urn and remove ball number 5. For a successful sequence, the next ball we remove must be either 4 or 6. Let's say it's 6. Now our sequence is 5, 6. Now we remove another ball, and this time the ball must be either a 4 or a 7. If it's ball number 4, then our sequence is 4, 5, 6. To continue a successful sequence, the next ball we remove must be either a 3 or a 7, and so on. At the end of our successful process, we will have 1, 2, 3, ..., n.

A good way to solve this problem is to *work backwards*. Let S_k be a set of numbered balls already drawn from the urn that is a "successful" sequence of k positive integers. Now start at the

end of the process, with $S_n = \{1, 2, 3, ..., n\}$ and work backwards.
There are two choices to obtain S_{n-1} from S_n. Remove either
ball 1 or ball n from the set S_n. There are two balls that we
could successfully remove (1 or n) and there are n balls in S_n, so
the probability of this first successful step is $2/n$. In the next
step, we will again have two successful choices and there will be
$n-1$ balls in the urn, so the probability for this step being
successful is $2/(n-1)$. We continue in this way until only one
ball remains in the urn. This is set S_1, and there is just one ball to
remove. So the final probability of success is $1/1$.

The backward process has the same probability as the forward
process, because the number of possibilities for S_k in the
backward process is equal to the number of possibilities for S_k in
the forward process.

The total probability is just the product of all the individual,
independent probabilities:

$$\frac{2}{n} \cdot \frac{2}{(n-1)} \cdot \frac{2}{(n-2)} \cdot \; ... \; \cdot \frac{2}{3} \cdot \frac{2}{2} \cdot \frac{1}{1} = \frac{2^{n-1}}{n!}. \quad \square$$

Probability problems are typically counter-intuitive, and they
can be quite frightening. A good approach is to start with a
small, finite set, say for $n = 5$, and work through the process one
step at a time.

Problem 8

Minimum Total Distance

Six people live along a street. Thinking of the street as a number line, their locations are at $x = 0, 15, 35, 60, 90$, and 120. Engineers want to build a bus stop at a place where the total distance the people must walk, to get to the bus stop from their homes, is the minimum possible total distance. Where should the bus stop be located on the x axis, and what is the minimum total distance?

Let's draw a picture to illustrate the problem.

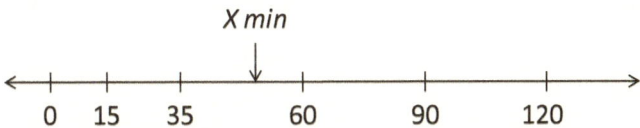

Point x_{min} will be the location of the bus stop. Right now, we don't know where x_{min} should be. We must figure that out.

If x_{min} were in the location shown in the figure, then we would have $x_{min} - 35 > 0$ and $x_{min} - 60 < 0$. To avoid having to worry about positives and negatives, let's just use absolute values to represent the distances that people must walk. You may remember the definition of *absolute value*: For real y, $|y| = +y$ if $y \geq 0$, and $|y| = -y$ if $y < 0$. Then, the person at point $x = 35$ would have to walk a distance of $|x_{min} - 35|$, and the person at point $x = 60$ would have to walk a distance of $|x_{min} - 60|$. This notation works because $|x - y| = |y - x|$.

We can now describe our problem by saying that the total distance the people must walk to get to the bus stop is

$$d_{tot}(x) = |x_{min}| + |x_{min} - 15| + |x_{min} - 35| + |x_{min} - 60| + |x_{min} - 90| + |x_{min} - 120|$$

We must find the value x_{min} that minimizes $d_{tot}(x)$.

If you don't know how to solve equations with absolute values, this can be an intimidating problem. What is a brute-force-dumb approach? One way is to try plugging different values of x into the equation and see what you get. Here are a few values: $d_{tot}(-5) = 350$, $d_{tot}(0) = 320$, $d_{tot}(17) = 256$, $d_{tot}(38) = 220$, $d_{tot}(59) = 220$, and $d_{tot}(70) = 240$. If you continue plugging in different values of x, after a while you will realize that $d_{tot}(x)$ *appears* to be minimized at a value of $d_{tot}(x_{min}) = 220$ whenever x_{min} lies *anywhere* in the range $35 \leq x_{min} \leq 60$. Now that we know the apparent answer to our problem, our task is to *prove* these conjectures. How can we do that?

A good place to start looking for a proof is to see if there are any *theorems* that resemble the equation for $d_{tot}(x)$. Do you know of any theorems involving a *sum of absolute values that must be minimized*? If not, you can look for such a theorem in a reference book. There is a famous and important theorem that looks like the thing we need. It is the *triangle inequality:*

Theorem 8.1. *For real numbers* x_1, x_2, ..., x_n,
$$|x_1|+|x_2|+...+|x_n| \geq |x_1 + x_2 + ...+ x_n|, \text{ with equality if and only if}$$
all the x_i *values have the same sign.*

Now, if we choose $x_{\min}$ so that it lies in the range $35 \leq x_{\min} \leq 60$, we can use the triangle inequality and cleverly arrange the x_i values so that all x_i are positive (have the same sign) and $x_{\min}$ cancels out of the inequality:

$$
\begin{aligned}
d_{tot}(x) &= |x_{\min}| + |x_{\min} - 15| + |x_{\min} - 35| + |x_{\min} - 60| + |x_{\min} - 90| + |x_{\min} - 120| \\
&= |x_{\min}| + |x_{\min} - 15| + |x_{\min} - 35| + |60 - x_{\min}| + |90 - x_{\min}| + |120 - x_{\min}| \\
&\geq |x_{\min} + x_{\min} - 15 + x_{\min} - 35 + 60 - x_{\min} + 90 - x_{\min} + 120 - x_{\min}| \\
&= |3x_{\min} - 3x_{\min} + 220| \\
&= |220| \\
&= 220. \ \square
\end{aligned}
$$

Using the triangle inequality, we have shown that if the bus stop is located anywhere in the range $35 \leq x_{min} \leq 60$, then the total distance the people must walk to get to the bus stop is minimized, and that minimum total distance is 220. (We haven't specified any units, but it could be, for example, 220 meters.)

Notice how we solved this problem. We used a combination of drawing a picture, choosing a suitable mathematical representation, experimenting with numbers to find a brute-force answer, and trying to find a theorem that resembles the situation in our problem. Then we massaged, or cleverly shaped, the theorem so that the minimizing condition in the triangle inequality would actually occur. *Good problem solvers are flexible problem solvers.*

Problem 9

Positive Integer Divisors

How many positive integer divisors does the number n have, where $n = 1007021035035021007001$?

This is a big number, so let's first look at a smaller number, say 10. The positive integer divisors of 10 are 1, 2, 5, and 10. So, there are four positive integer divisors of 10. How can we find the number of positive integer divisors of an arbitrary positive integer? The key is to first prime factorize the number. Suppose $m = 1800$. The prime factorization of m is $m = 1800 = 2^3 \cdot 3^2 \cdot 5^2$. Any divisor d of m must have the form $2^a \cdot 3^b \cdot 5^c$, where $0 \le a \le 3$, $0 \le b \le 2$, and $0 \le c \le 2$. If all of a, b, c are zero, then $2^0 \cdot 3^0 \cdot 5^0 = 1$, and 1 is also a divisor of m. Thus, we have 4 choices for a (i.e., 0, 1, 2, or 3), 3 choices for b (i.e., 0, 1, or 2), and 3 choices for c (i.e., 0, 1, or 2). The total number of positive integer divisors of 1800 is $4 \cdot 3 \cdot 3 = 36$.

In general, if $m = p_1^{a_1} \cdot p_2^{a_2} \cdot ... \cdot p_k^{a_k}$, for distinct primes $p_1, p_2, ... , p_k$, then the number of positive integer divisors of m is $(a_1 + 1)(a_2 + 1)...(a_k + 1)$.

Returning to the number n in the problem, one way to find the number of its positive integer divisors is to prime factorize n. Then use the preceding method. The difficulty with this approach is that n is a big number, and it may be difficult to find all the prime factors of n. Prime factorization is a difficult problem in mathematics. We need to find a shortcut.

The number n has a bunch of zeros in it, so maybe we can break the number into "pieces," along the zero boundaries, and add them up like this:

$$
\begin{aligned}
n = {} & 1007021035035021007001 \\
= {} & 1000000000000000000000 \\
& + 7000000000000000000 \\
& + 21000000000000000 \\
& + 35000000000000 \\
& + 35000000000 \\
& + 21000000 \\
& + 7000 \\
& + 1
\end{aligned}
$$

I don't like all those zeros, since it's hard to see what's going on, so let's use scientific notation. We then have the following:

$$n = 1 \cdot 10^{21} + 7 \cdot 10^{18} + 21 \cdot 10^{15} + 35 \cdot 10^{12} + 35 \cdot 10^{9} + 21 \cdot 10^{6} + 7 \cdot 10^{3} + 1 \cdot 10^{0}$$

Now we can see a pattern with the powers of 10. The exponents decrease by three each time (21, 18, 15, 12, 9, 6, 3, 0). What about the numbers 1, 7, 21, 35, 35, 21, 7, 1? What do they mean? The secret to solving this problem depends on us understanding this sequence. Have you seen it before?—maybe not. However, one of our basic problem solving principles is to *always look for your numbers in Pascal's triangle* (see *Chapter 4*). Indeed, the unknown sequence is, in fact, the 7th row of Pascal's triangle. (The first row is counted as row zero.)

$$
\begin{array}{ccccccccccccccc}
& & & & & & & 1 & & & & & & & \\
& & & & & & 1 & & 1 & & & & & & \\
& & & & & 1 & & 2 & & 1 & & & & & \\
& & & & 1 & & 3 & & 3 & & 1 & & & & \\
& & & 1 & & 4 & & 6 & & 4 & & 1 & & & \\
& & 1 & & 5 & & 10 & & 10 & & 5 & & 1 & & \\
& 1 & & 6 & & 15 & & 20 & & 15 & & 6 & & 1 & \\
1 & & 7 & & 21 & & 35 & & 35 & & 21 & & 7 & & 1
\end{array}
$$

The elements in the *k*th row of Pascal's triangle are just the *binomial coefficients*:

$$
\binom{k}{r} = \frac{k!}{r!(k-r)!}, \text{ for } 0 \le r \le k.
$$

This means that our sequence 1, 7, 21, 35, 35, 21, 7, 1 is really the sequence $\binom{7}{0}, \binom{7}{1}, \binom{7}{2}, \binom{7}{3}, \binom{7}{4}, \binom{7}{5}, \binom{7}{6}, \binom{7}{7}$. So, the number n can be written as

$$n = \binom{7}{0} \cdot 10^{21} + \binom{7}{1} \cdot 10^{18} + \binom{7}{2} \cdot 10^{15} + \binom{7}{3} \cdot 10^{12} + \binom{7}{4} \cdot 10^{9}$$
$$+ \binom{7}{5} \cdot 10^{6} + \binom{7}{6} \cdot 10^{3} + \binom{7}{7} \cdot 10^{0}$$

$$= \binom{7}{0} \cdot (10^3)^7 + \binom{7}{1} \cdot (10^3)^6 + \binom{7}{2} \cdot (10^3)^5 + \binom{7}{3} \cdot (10^3)^4 + \binom{7}{4} \cdot (10^3)^3$$
$$+ \binom{7}{5} \cdot (10^3)^2 + \binom{7}{6} \cdot (10^3)^1 + \binom{7}{7} \cdot (10^3)^0$$

$$= (10^3 + 1)^7.$$

The last sneaky step came from the *binomial theorem*. The binomial theorem is so important that you should memorize it.

Theorem 9.1. $(a+b)^n = \sum_{k=0}^{n} \binom{n}{k} a^k b^{n-k}$.

Using the binomial theorem, with $a = 10^3$, $b = 1$, and $n = 7$, the sum of all those binomial coefficients reduces to simply $n = (10^3 + 1)^7 = 1001^7$. Now our original problem is much easier.

We have discovered that $n = 1007021035035021007001 =$ $(1001)^7$. We just need to prime factorize 1001. We get, possibly by trial and error, $1001 = 7 \cdot 11 \cdot 13$. So, $n = (7 \cdot 11 \cdot 13)^7 =$ $7^7 \cdot 11^7 \cdot 13^7$. Therefore, the number of positive integer divisors of n is $(7+1)(7+1)(7+1) = 8^3 = 512$. There are 512 positive integer divisors of 1007021035035021007001. □

Problem 10

Calculator Error

Evaluate the expression $N = (1 - 2.2 \times 10^{-22})^{2.2 \times 10^{22}}$.

This problem looks easy, but it illustrates the danger of thoughtlessly using a calculator or computer. If you plug the expression into a calculator, most calculators give the wrong answer: 1. Even worse, many symbolic algebra computer programs also give the wrong answer. The answer is not 1. This is a case where blind faith in technology can lead you astray. But why?

The crux of the problem is that we have both a large number (2.2×10^{22}) and a small number (2.2×10^{-22}) in the same expression, and those two numbers work against each other. To put things in perspective, the number of molecules in one liter of ideal gas at 25 C and 1 atmosphere pressure is 6.02×10^{23}. In our problem, we have the astronomically large number 2.2×10^{22} along with the infinitesimally small number 2.2×10^{-22}. Most calculators can't handle this. How can we approach this problem?

One idea is to gradually approach our number N and see what happens. Let's replace N by a function of n, where n is a positive integer. We then have

$$N(n) = (1 - 2.2 \times 10^{-n})^{2.2 \times 10^n}.$$

Now just plug some small values of n into the function $N(n)$ and see what happens. The calculator can handle this problem. Here's what we get:

$$N(1) = 0.0042274771...$$
$$N(2) = 0.0074911424...$$
$$N(3) = 0.0078650072...$$
$$N(4) = 0.0079028448...$$
$$N(5) = 0.0079066331...$$
$$N(6) = 0.0079070119...$$

It certainly appears, from the numerical data, that as n gets larger $N(n)$ approaches approximately 0.0079. This is what mathematicians call a *limiting* process, and it is a central theme in calculus. We won't worry about calculus. We are interested in proving the following conjecture:

Conjecture. $\lim_{n \to 22} (1 - 2.2 \times 10^{-n})^{2.2 \times 10^n} = 0.0079...$.

As always, we look at our problem and ask if we can think of any mathematical theorems that resemble our situation. This is where it helps to have some mathematical knowledge. If you really want to become a good mathematical problem solver, you can't escape the necessity of knowing some mathematics. You will need to know some important theorems. If you don't know the right theorems, then you will have to look for them in a good reference book. Appendix B is a good place to start.

Is there a theorem that looks something like our problem? Indeed, there is. Here it is:

Theorem 10.1. $\displaystyle\lim_{x \to \infty}\left(1 - \frac{1}{x}\right)^x = \frac{1}{e} \approx 0.367879\ldots$.

The number e, called *Euler's number*, is approximately 2.71828.... It's an irrational number. This number is really important, and it shows up everywhere in mathematics. Now our problem is to somehow massage the number N, given in the problem, so that it looks like Theorem 10.1. With a little number juggling, here is how we can do that:

$$N = (1 - 2.2 \times 10^{-22})^{2.2 \times 10^{22}}$$

$$= \left(1 - \frac{1}{0.4545 \times 10^{22}}\right)^{2.2 \times 10^{22}}$$

$$= \left(\left(1 - \frac{1}{0.4545 \times 10^{22}}\right)^{0.4545 \times 10^{22}}\right)^{4.84}$$

$$\approx \left(\frac{1}{e}\right)^{4.84}$$

$$= 0.00791....$$

Instead of taking the limit as x tends to infinity, we just used a very large value of x, namely $x = 0.4545 \times 10^{22}$. Then, by Theorem 10.1, we realize that $\left(1 - \dfrac{1}{0.4545 \times 10^{22}}\right)^{0.4545 \times 10^{22}}$ is approximately $\dfrac{1}{e}$. This is how mathematicians approximate things. *Estimating* and *approximating* are useful problem solving skills. Just to be complete, here is another theorem that is very useful in doing mathematical approximations:

Theorem 10.2. $\lim\limits_{x \to \infty} \left(1 + \dfrac{1}{x}\right)^{x} = e \approx 2.71828....$.

Try using Theorems 10.1 and 10.2 for your estimating problems when they involve either very large or very small numbers.

Problem 11

Root Canal

Find the largest possible real value of the function $f(x, y)$ where

$$f(x, y) = \sqrt{(x-20)(y-x)} + \sqrt{(140-y)(20-x)} + \sqrt{(x-y)(y-140)}$$

such that $-40 \le x \le 100$ and $-20 \le y \le 200$.

This problem looks as painful as a dental root canal. How can we begin to explore this problem? A natural idea is to square both sides of the equation in the hope that we can get rid of those nasty square roots. This approach might work, but the algebra gets messy. Is there a better way? Let's go back to basics.

What is the problem statement asking us to do? We want to find the largest possible *real* value of $f(x, y)$ for certain constraints on x and y. Those are the key words: "real value." For $f(x, y)$ to be a real number, we cannot have a square root of a negative number. Why? The square root of a negative number is imaginary. Think of $\sqrt{-5}$. That's not a real number. There is no real number whose square is -5. So, a key insight is that we cannot take the square root of a negative number if $f(x, y)$ is real. That means that each of the radicands must be positive or zero:

$$(x-20)(y-x) \geq 0,$$
$$(140-y)(20-x) \geq 0,$$
$$(x-y)(y-140) \geq 0.$$

Now what should we do? We have two inequalities that contain either $y-x$ or $x-y$. Let's try to get those expressions by themselves. Why? Our intuition tells us that it might be important to know *how x* is related to y. If we knew that, then maybe we could eliminate x (or y) and simplify $f(x, y)$. Let's do that.

From the first inequality, we have $(x-20)(y-x) \geq 0$. As long as $x-20 \neq 0$, we can divide through by $x-20$ to get $y-x \geq 0$, or $y \geq x$. From the third inequality, we have $(x-y)(y-140) \geq 0$. So long as $y-140 \neq 0$, we can divide through by $y-140$ to get $x-y \geq 0$, or $x \geq y$. Thus, we see that $y \geq x$ if $x \neq 20$ and $x \geq y$ if $y \neq 140$. But wait a minute! We have shown that $y \leq x \leq y$ provided that $x \neq 20$ and $y \neq 140$. This means that $x = y$ for $x \neq 20$ and $y \neq 140$.

Now, if $x = y$, then $f(x, y)$ becomes

$$\begin{aligned} f(x, y) &= \sqrt{(x-20)(x-x)} + \sqrt{(140-x)(20-x)} + \sqrt{(x-x)(x-140)} \\ &= \sqrt{0} + \sqrt{(140-x)(20-x)} + \sqrt{0} \\ &= \sqrt{(140-x)(20-x)}. \end{aligned}$$

Squaring both sides, we get $f^2(x, y) = (140 - x)(20 - x)$

$= x^2 - 160x + 2800$. Don't forget that x cannot be 20 or 140. With $x = y$ the given constraints become $-40 \leq x \leq 100$ and $-20 \leq x \leq 200$. These constraints must both be true, and they overlap. So, we only need to examine the overlap: $-20 \leq x \leq 100$. Thus, our problem is now to find the value of x that maximizes the value of $x^2 - 160x + 2800$. This polynomial becomes larger as x becomes more negative. The most negative that x could become, on the interval $-20 \leq x \leq 100$, is $x = -20$.

Now, if we substitute $x = y = -20$ into $f(x, y)$, in the problem statement, we get the largest possible real value for $f(x, y)$: $f_{max}(x, y) = f(-20, -20) = 80$. $\square$

Remember, constraints are your friend. Mathematical constraints may seem like a nuisance, but they often help you solve your problem by restricting what is possible.

Problem 12

Two, Three, Five

Find the smallest positive integer n such that $n/2$ is a perfect square, $n/3$ is a perfect cube, and $n/5$ is a perfect fifth power.

When you read a problem statement that tells you to find something, there is a natural tendency to try to find it. You may, for example, get a long list of positive integers and look for a number in the list that satisfies the given conditions. That would be a brute-force-dumb solution. Just do a systematic exhaustive search. You could even write a computer program to conduct the search for you while you are away at lunch. The trouble with that method is that we have no idea how long we will need to search for our special number. The number we are looking for could be astronomically large. There must be a better way.

Here is a tip: Whenever you see the word "find," replace it with the word "build." Instead of trying to *find* the required object, see if you can *build* the object. This approach gives you a lot more control over the problem. How can we build the required number for this problem?

The statement that $n/2$ is a perfect square, $n/3$ is a perfect cube, and $n/5$ is a perfect fifth power, suggests that n is divisible by 2, 3, and 5. We might imagine that n is divisible by other prime numbers, but the problem asks us to find the *smallest* such number n. So, there is no need to make the number larger by adding extra factors. Therefore, we can assume that the number n has the form $n = 2^a \cdot 3^b \cdot 5^c$, for some nonnegative integers a, b, and c. Now we just need to find the smallest values of a, b, and c that fit the problem statement.

We see that $n/2 = 2^{a-1} \cdot 3^b \cdot 5^c$. Also, $n/3 = 2^a \cdot 3^{b-1} \cdot 5^c$ and $n/5 = 2^a \cdot 3^b \cdot 5^{c-1}$. Since $n/2$ must be a perfect square, $a-1$, b, and c must all be even. Since $n/3$ is a perfect cube, a, $b-1$, and c must all be multiples of 3. And since $n/5$ is a perfect fifth power, a, b, and $c-1$ must all be multiples of 5.

Let's rephrase this new requirement. We must have a be a multiple of 3 and 5, with $a - 1$ being a multiple of 2. We can satisfy this condition if $a = 15$. Next, b must be a multiple of 2 and 5, with $b - 1$ being a multiple of 3. If $b = 10$, that works. Finally, c must be a multiple of 2 and 3, with $c - 1$ being a multiple of 5. Here, $c = 6$ works. So, we have $a = 15$, $b = 10$, and $c = 6$. The number we seek is $n = 2^{15} \cdot 3^{10} \cdot 5^6 =$ 30,233,088,000,000, or 30 trillion 233 billion 88 million. Now that's a big number, on the order of the National Debt. □

Problem 13

Sum of Five Squares

Prove that the sum of the squares of five consecutive integers cannot be a perfect square.

This problem asks us to prove something. A good way to approach problems like this is to assume that the premise is possible, and then show that this assumption results in a contradiction. The contradiction is some mathematical conclusion that cannot possibly be true. This kind of proof—*proof by contradiction*—was discussed in Chapter 6.

Let's assume that the sum of the squares of five consecutive integers is a perfect square. How can we write this condition mathematically? If n is an integer, then five consecutive integers are n, $n + 1$, $n + 2$, $n + 3$, and $n + 4$. The sum of squares of these five consecutive integers is $n^2 + (n+1)^2 + (n+2)^2 + (n+3)^2 + (n+4)^2 = 5n^2 + 20n + 30$. We could work with this expression, and there is nothing wrong with that. But let's not get too attached to one approach. Let's continue to play around with a representation for five consecutive squares. Another way to represent five consecutive integers is $n - 2$, $n - 1$, n, $n + 1$, and

$n+2$. This representation is symmetrical about n, and maybe some of the pluses and minuses will cancel each other out. Let's see what happens. We have $(n-2)^2+(n-1)^2+n^2+(n+1)^2+(n+2)^2=5n^2+10$. This is indeed a simpler expression than $5n^2+20n+30$, so let's work with this new expression.

Assuming that the sum of the squares of five consecutive integers is a perfect square, we have, for some positive integer m:

$$(n-2)^2+(n-1)^2+n^2+(n+1)^2+(n+2)^2=5n^2+10=m^2.$$

Factoring out 5, we get $5(n^2+2)=m^2$. Note that 5 is a prime number. Euclid's Lemma tells us that if a prime divides a product, then it must divide at least one factor. Since 5 divides the left-hand side of $5(n^2+2)=m^2$, 5 must divide m^2. And, by Euclid's Lemma, that means that 5 divides m. So, $m/5=r$, for some integer r, and we have $n^2+2=m\cdot r$. Also, since m is a multiple of 5, $m\cdot r$ is a multiple of 5. This tells us that n^2+2 is a multiple of 5. Using modular arithmetic (see discussion below), we can write this as $n^2+2\equiv 0\ (\mathrm{mod}\,5)$, or $n^2\equiv -2\ (\mathrm{mod}\,5)$, or $n^2\equiv 3\ (\mathrm{mod}\,5)$. Now we just need to show that this last condition is impossible.

Every integer n is congruent to one of the residues (integers) $\{0, 1, 2, 3, 4\}$. So, we can plug each one of these into n^2, reduce it modulo 5, and see what we get:

128

$$0^2 \equiv 0 \pmod 5$$
$$1^2 \equiv 1 \pmod 5$$
$$2^2 \equiv 4 \pmod 5$$
$$3^2 \equiv 9 \equiv 4 \pmod 5$$
$$4^2 \equiv 16 \equiv 1 \pmod 5$$

None of these is congruent to 3 modulo 5 as required by $n^2 \equiv 3 \pmod 5$. Therefore, it is impossible. The sum of the squares of five consecutive integers cannot be a perfect square.

In this problem we used modular arithmetic. Modular arithmetic is an indispensable tool for solving problems in number theory. You could still solve the problem without using modular arithmetic, but it would be more difficult. You would have to use a "divide-and-conquer" approach. We would consider five separate cases for n: $n = 5k$, $n = 5k+1$, $n = 5k+2$, $n = 5k+3$, and $n = 5k+4$, for some integer k. We don't need to consider any other cases. Why? Suppose, for example, that $n = 5k+6$. But $5k+6 = 5k+5+1 = 5(k+1)+1 = 5r+1$. This still has the *form* of $5k+1$ with k replaced by an integer r. You would then substitute these five cases into $5n^2 + 10 = m^2$ and show that contradictions result. It is much easier to solve this problem using modular arithmetic.

Modular arithmetic, modulo m, is just like telling time on an m-hour clock. Arithmetic modulo 12 is like telling time on a standard 12-hour clock. We would write 7 o'clock as

7 (mod 12), which we read as "seven modulo twelve." A time like 14 o'clock, which is the same as 2 pm, can be written as 14 (mod 12) $\equiv$ 2 (mod 12), or "14 is congruent to 2 modulo 12." And 12 o'clock is the same as 0 o'clock, so $12 \equiv 0$ (mod 12).

Basically, if $a \equiv b$ (mod m), then $a - b$ is divisible by m. We can write this as $a - b \equiv 0$ (mod m). The basic properties of modular arithmetic are the following. For integers a, b, c, k, and positive integer m, we have:

1. If $a \equiv b$ (mod m), then $a - b \equiv 0$ (mod m).
2. If $a \equiv b$ (mod m), then $a + c \equiv b + c$ (mod m).
3. If $a \equiv b$ (mod m), then $a \cdot k \equiv b \cdot k$ (mod m).
4. If $a \equiv b$ (mod m) and $b \equiv c$ (mod m), then $a \equiv c$ (mod m).
5. If $a \equiv b$ (mod m), then $b \equiv a$ (mod m).

You can try proving these from the basic definition that if $a \equiv b$ (mod m), then $a - b$ is divisible by m. Be careful, though! In general, cancellation does *not* work for modular arithmetic. Generally, if $a \cdot k \equiv b \cdot k$ (mod m), it is NOT true that $a \equiv b$ (mod m). Cancellation only works if k and m are coprime, or have no common prime factor.

Problem 14

Find the Minimum

Let x, y, z, and t be positive real numbers such that

$x+y+z+t=1$. What is the minimum value of $\dfrac{1}{x}+\dfrac{1}{y}+\dfrac{4}{z}+\dfrac{16}{t}$?

If you have never seen a problem like this, it can be a bit scary. As usual, let's get comfortable with the problem by plugging in some numbers. We have the constraints that x, y, z, and t must be positive real numbers, and $x+y+z+t=1$. We

want to minimize the function $f(x,\ y,\ z,\ t)=\dfrac{1}{x}+\dfrac{1}{y}+\dfrac{4}{z}+\dfrac{16}{t}$.

We can start with small cases and special cases. First, note that $f(0, 0, 0, 1)$ is undefined because division by zero is undefined. So, none of x, y, z, t can be zero. What about the special case where $x=y=z=t=1/4$, or 0.25? In this case, we certainly have $x+y+z+t=1$, and $f(1/4, 1/4, 1/4, 1/4)=88$. So, the minimum value of $f(x, y, z, t)$ is less than or equal to 88. Can we do better? Try some more numbers and see what happens. In $f(x, y, z, t)$, the term $16/t$ dominates the expression. So, if we want $f(x, y, z, t)$ to be a minimum, then t should be larger, not smaller. We also have symmetry with x and y, so they should be equal. We can try $f(1/8, 1/8, 1/4, 1/2) = 64$. This is better than 88. If you continue to play around with different combinations of numbers for x, y, z, and t, you will find that it seems impossible to do better than 64 as a minimum for $f(x, y, z, t)$. Is 64 the required minimum? How can we know?

It's challenging to find a good, rigorous solution to this problem without some mathematical knowledge. But what mathematical knowledge do we need? How can we possibly know which mathematical theorems, if any, will help us solve the problem?

There are some subtle clues in the problem statement. An important clue is that this problem asks us to find the *minimum* of some function. Problems involving *optima* (maxima or minima) are generally solved in one of two ways. Either we use calculus or we use inequalities. Calculus certainly works, and there are methods, like the method of Lagrange Multipliers, that are general and powerful tools for solving optimization problems. More likely than not, however, we can solve this problem without calculus. Why? Because Olympiad mathematics problems almost never require calculus. This is a kind of "meta-knowledge" about the nature of the problem and the psychology of the person who crafted the problem. Meta-knowledge is good. Use it whenever you can. This can only mean one thing: We must solve this problem using an inequality. But which inequality?

In mathematics, there are many inequalities, and new ones are being discovered every day. But only a handful of inequalities are frequently used in day-to-day problem solving. These are the Arithmetic-Geometric-Harmonic Mean Inequality and the Cauchy-Schwartz Inequality. There are a few other important inequalities, but these are the "heavy lifters" of mathematical problem solving. Let's first review these two inequalities. Then we will look for clues in our problem statement to decide which inequality is best.

Theorem 14.1 (AGH Inequality). *The arithmetic mean is greater than or equal to the geometric mean, which is greater than or equal to the harmonic mean.* More precisely, for any n positive real numbers $a_1, a_2, ..., a_n$,

$$\frac{a_1 + a_2 + ... + a_n}{n} \geq (a_1 \cdot a_2 \cdot ... \cdot a_n)^{1/n} \geq \frac{n}{\dfrac{1}{a_1} + \dfrac{1}{a_2} + ... + \dfrac{1}{a_n}},$$

with equality if $a_1 = a_2 = ... = a_n$.

Theorem 14.2 (Cauchy-Schwartz Inequality). *The square of a sum of products is less than or equal to the product of the sums of squares.* More precisely, for arbitrary real numbers $a_1, a_2, ..., a_n$ and $b_1, b_2, ..., b_n$,

$$(a_1 b_1 + ... + a_n b_n)^2 \leq (a_1^2 + ... + a_n^2)(b_1^2 + ... + b_n^2),$$

with equality if $(a_1, \ldots, a_n)$ and $(b_1, \ldots, b_n)$ are proportional.

Returning to our problem, which inequality should we use? The key observation is the presence of the numbers 1, 1, 4, and 16 in $f(x, y, z, t) = \dfrac{1}{x} + \dfrac{1}{y} + \dfrac{4}{z} + \dfrac{16}{t}$. These are all perfect squares: $1 = 1^2$, $4 = 2^2$, and $16 = 4^2$. Which inequality uses the word "square"? It's the Cauchy-Schwartz Inequality. Now we just need to play around with the Cauchy-Schwartz Inequality and try to force-fit it into our problem. This may take some experimentation, but here is how we can make it work.

Think of $f(x, y, z, t)$ as

$$f(x, y, z, t) = \dfrac{1^2}{(\sqrt{x})^2} + \dfrac{1^2}{(\sqrt{y})^2} + \dfrac{2^2}{(\sqrt{z})^2} + \dfrac{4^2}{(\sqrt{t})^2}.$$ By the Cauchy-

Schwartz Inequality, we have

$$\left(\dfrac{1}{\sqrt{x}} \cdot \sqrt{x} + \dfrac{1}{\sqrt{y}} \cdot \sqrt{y} + \dfrac{2}{\sqrt{z}} \cdot \sqrt{z} + \dfrac{4}{\sqrt{t}} \cdot \sqrt{t} \right)^2 \le$$

$$\left(\left(\dfrac{1}{\sqrt{x}}\right)^2 + \left(\dfrac{1}{\sqrt{y}}\right)^2 + \left(\dfrac{2}{\sqrt{z}}\right)^2 + \left(\dfrac{4}{\sqrt{t}}\right)^2 \right) \left((\sqrt{x})^2 + (\sqrt{y})^2 + (\sqrt{z})^2 + (\sqrt{t})^2 \right)$$

$$(1+1+2+4)^2 \le \left(\frac{1^2}{x} + \frac{1^2}{y} + \frac{2^2}{z} + \frac{4^2}{t} \right)(x+y+z+t)$$

$$= \left(\frac{1^2}{x} + \frac{1^2}{y} + \frac{2^2}{z} + \frac{4^2}{t} \right)$$

In the last step, we used the given constraint that $x+y+z+t=1$. Finally, we have

$$64 \le \frac{1}{x} + \frac{1}{y} + \frac{4}{z} + \frac{16}{t}.$$

We have just one final step before we can conclude that the minimum value of $f(x, y, z, t) = \frac{1}{x} + \frac{1}{y} + \frac{4}{z} + \frac{16}{t}$ is 64. We must show that the lower bound of 64 can actually be achieved. But wait! We already did that. In our numerical investigations, we showed that $f(1/8, 1/8, 1/4, 1/2) = 64$. We are done. □

Problem 15

Least Value

Find the least value of $f(x, y, z) = \dfrac{x}{y} + \dfrac{3y}{z} + \dfrac{9z}{x}$ for positive real

numbers x, y, and z.

After solving Problem 14, this problem should be less frightening. Since it is a minimization problem, we know that we will either use calculus or inequalities to solve it. Of course, we will be using an inequality for this problem.

Ordinarily, there is no guarantee that a function has a least value on some domain. Some functions, like $y = \log x$, have no least value for positive real x. With this caution in mind, we can plug some numbers into $f(x, y, z)$ to get a feel for how this function behaves. First, note that none of x, y, or z can equal zero. This is because division by zero is undefined. That's OK, because the problem statement tells us that x, y, and z must be positive real numbers. So, each of x, y, and z is greater than zero. Let's try a few calculations for the special case $x = y = z$. We have, for example, $f(1, 1, 1) = (1/1) + 3 \cdot (1/1) + 9 \cdot (1/1) = 13$. Similarly, we have $f(0.5, 0.5, 0.5) = 13$. Can we do any better?

Maybe we can do better if x, y, and z are not all equal. For example, $f(2, 1, 1) = 9.5$ and $f(3, 1, 1) = 9$. You can play around with different combinations of x, y, and z. But it looks like the minimum, assuming it exists, is around 9. Now we have a feel for the problem.

If the minimum is 9, that's another hint that we should use an inequality to solve this problem. Why? The function $f(x, y, z)$ has positive integers in it, like 1, 3, and 9. It's very suspicious that the minimum might also be a positive integer, like 9, which is itself a multiple of 3. Some combination of 1, 3, and 9 is giving rise to 9, and this is clear evidence that the math god is at work. Patterns like this are clues, and we should learn to recognize little clues in our problems.

To solve this problem using an inequality, we need to make an educated guess about which inequality is likely to work. The function $f(x, y, z)$ contains the numbers 1, 3, and 9. Although 1 and 9 are squares, 3 is not a perfect square. This observation does not rule out the Cauchy-Schwartz Inequality, and it might work, but the Arithmetic-Geometric Mean Inequality (Theorem 14.1, Problem 14) might be a better bet. Another reason why the Arithmetic-Geometric Mean Inequality might work better for this problem is because when we multiply x/y, $3y/z$, and $9z/x$, all the x's, y's, and z's cancel out.

By Theorem 14.1, the arithmetic mean is greater than or equal to the geometric mean, with equality if all the terms are equal. This gives us the following:

$$\frac{1}{3}\left(\frac{x}{y}+\frac{3y}{z}+\frac{9z}{x}\right) \geq \left(\frac{x}{y}\cdot\frac{3y}{z}\cdot\frac{9z}{x}\right)^{1/3} = (27)^{1/3} = 3.$$

This means that $\frac{x}{y}+\frac{3y}{z}+\frac{9z}{x} \geq 9$. We already saw, when we did some numerical calculations, that $f(3, 1, 1) = 9$. So 9 is the least value of $f(x, y, z)$ for positive real numbers x, y, z. That's the answer to the problem. □

Suppose that our experimental testing did not reveal that $f(3, 1, 1)$ gives the minimum value of $f(x, y, z)$. How could we find the values of x, y, and z that give the minimum (assuming that the minimum is actually achievable)? Here is one way to analyze this problem.

The Arithmetic-Geometric Mean Inequality tells us that equality is achieved when all the terms are equal. This means that we want $\frac{x}{y} = \frac{3y}{z} = \frac{9z}{x} = r$, for some positive real number r.

We also know that $r^3 = \frac{x}{y}\cdot\frac{3y}{z}\cdot\frac{9z}{x} = 27$, so $r = 3$. This gives us three equations to solve: $\frac{x}{y} = 3$, $\frac{3y}{z} = 3$, and $\frac{9z}{x} = 3$. There are

139

infinitely many solutions to these equations. For one solution, we can let $x = 3$, $y = 1$, and $z = 1$. Then we get $f(3, 1, 1) = 9$, and we have our desired minimum.

Optimization problems usually ask us to find a maximum or minimum of something given some constraints. These problems can usually be solved with advanced methods from calculus (e.g., the method of Lagrange Multipliers). But they can often be solved elegantly using inequalities, as we did in this problem.

There are a few cautions to keep in mind whenever you solve an optimization problem. First, you must prove that an optimum (maximum or minimum) actually exists. Second, you must verify that your answer satisfies all of the given constraints. Finally, bear in mind that the optimum—if it exists—may not be unique. There may be several values of, say x, y, and z, that give the optimum result.

Problem 16

A Classic Sum

Write the sum $\dfrac{1}{1\cdot 2}+\dfrac{1}{2\cdot 3}+\dfrac{1}{3\cdot 4}+...+\dfrac{1}{99\cdot 100}$ as a fraction in lowest terms.

Some mathematical problems are most easily solved if you know a special trick. Most mathematicians have a bag of special tricks that they have learned over the years. If you don't know the trick, you might not think of it. If you do know the trick, you can use it to solve a large class of similar problems. In this problem, we will begin by using our standard methods. Then we will solve the problem using a trick that often works for doing summations. This will prepare us for our next problem, Problem 17, where we will use the special summation trick on a rather difficult problem.

For notational convenience, let us refer to the given sum as $S(99)$. You might wonder, why not let the sum be $S(100)$? Well, you could. It's just a matter of preference. When you think of mathematics as a fine art, you realize that you can create your art however you want it to be. The sum in the problem has

99 terms, but it is usually better to generalize things and consider a more general sum as follows:

$$S(n) = \frac{1}{1 \cdot 2} + \frac{1}{2 \cdot 3} + \ldots + \frac{1}{n(n+1)}.$$

This problem could actually be solved using a simple brute-force-dumb method: We could just get a calculator and add up the 99 terms. That would work, but it's not very interesting. Let's see what else we can do.

If you look at small cases, you can calculate the first few partial sums: $S(1) = \frac{1}{1 \cdot 2} = \frac{1}{2}$, $S(2) = \frac{1}{1 \cdot 2} + \frac{1}{2 \cdot 3} = \frac{2}{3}$, and

$S(3) = \frac{1}{1 \cdot 2} + \frac{1}{2 \cdot 3} + \frac{1}{3 \cdot 4} = \frac{3}{4}$. From these few values of $S(n)$, it looks like a pattern is emerging. Our *conjecture* is that

$S(n) = \frac{n}{n+1}$. How can we prove this conjecture? One way is to use mathematical induction, which we discussed in Chapter 6.

We want to prove, by mathematical induction, that

$S(n) = \frac{1}{1 \cdot 2} + \frac{1}{2 \cdot 3} + \ldots + \frac{1}{n(n+1)} = \frac{n}{n+1}$. First, let's show that this

formula is true for the initial case of $n = 1$. We have

$S(1) = \frac{1}{1 \cdot 2} = \frac{1}{2}$, so the formula for $S(n)$ is certainly true when

$n = 1$. Next, we must show that *if* $S(n) = \frac{n}{n+1}$, *then*

$S(n+1) = \dfrac{n+1}{n+2}$. To do this, we can assume that $S(n) = \dfrac{n}{n+1}$,

then consider the sum for $S(n+1)$ and see if we actually get

$S(n+1) = \dfrac{n+1}{n+2}$. Here is the calculation. Try working through it

yourself.

$$S(n+1) = \frac{1}{1\cdot 2} + \frac{1}{2\cdot 3} + \ldots + \frac{1}{n(n+1)} + \frac{1}{(n+1)(n+2)}$$

$$= S(n) + \frac{1}{(n+1)(n+2)}$$

$$= \frac{n}{n+1} + \frac{1}{(n+1)(n+2)}$$

$$= \frac{1}{n+1}\left(n + \frac{1}{n+2}\right)$$

$$= \frac{1}{n+1}\left(\frac{n^2 + 2n + 1}{n+2}\right)$$

$$= \frac{1}{n+1}\left(\frac{(n+1)(n+1)}{n+2}\right)$$

$$= \frac{n+1}{n+2}. \quad \square$$

Now that we have proved that $S(n) = \dfrac{n}{n+1}$, we can immediately

solve our problem. The answer is $S(99) = \dfrac{99}{100}$, and this fraction

cannot be reduced any further.

Now that we did things the hard way, let's learn a little trick. This trick often works for a lot of summation problems, so it's good to know it. It's called the *telescoping sum* trick.

Suppose that we can write some sum, which we want to simplify, as $S(n) = \sum_{k=1}^{n-1} (f(k+1) - f(k))$, for some function $f(k)$ defined on the positive (or sometimes nonnegative) integers. Watch what happens when you add up the terms. Almost everything collapses, or "telescopes," and we are left with a simple result:

$$S(n) = \sum_{k=1}^{n-1} (f(k+1) - f(k))$$
$$= f(n) - f(n-1) + f(n-1) + f(n-2) + \ldots - f(2) + f(2) - f(1)$$
$$= f(n) - f(1).$$

That's amazing! For our problem, note that each term in the sum $S(n)$ has the form $\dfrac{1}{k(k+1)} = \dfrac{1}{k} - \dfrac{1}{k+1}$. So, for example,

$\dfrac{1}{3 \cdot 4} = \dfrac{1}{3} - \dfrac{1}{4}$. We could use mathematical symbols, but let's use numbers to see what's going on:

$$S(99) = \frac{1}{1\cdot 2} + \frac{1}{2\cdot 3} + \frac{1}{3\cdot 4} + \ldots + \frac{1}{99\cdot 100}$$

$$= \left(\frac{1}{1} - \frac{1}{2}\right) + \left(\frac{1}{2} - \frac{1}{3}\right) + \left(\frac{1}{3} - \frac{1}{4}\right) + \ldots + \left(\frac{1}{99} - \frac{1}{100}\right)$$

$$= 1 - \frac{1}{100}$$

$$= \frac{99}{100}.$$

The telescoping sum trick saves a lot of work—when we can use it successfully. The hard part with using the telescoping sum trick is that we must figure out how to represent the given sum, in our problem, in the form $S(n) = \sum_{k=1}^{n-1}(f(k+1) - f(k))$. If you can do that, your problem won't stand a chance against you.

The next problem, Problem 17, is much harder, but it nicely illustrates the full power of the telescoping sum trick.

Problem 17

Reciprocal of a Sum

Consider the sequence of positive integers $\{a_n\}$ defined by the recurrence relation $a_n = 2a_{n-1} + a_{n-2}$ for $n \geq 2$ and $a_0 = 1$, $a_1 = 213$. Let the infinite sum S be given by

$$S = \sum_{i=1}^{\infty} \frac{a_{i-1}}{a_i^2 - a_{i-1}^2}. \text{ Find the value of } 1/S.$$

One approach, to get a feel for this problem, is to plug some numbers into the sum S and see what happens. Since the sum S has infinitely many terms, we can actually calculate only the first several terms. But if S converges rather quickly, assuming that it does converge, the first several terms may give us some insight into the final value of the sum. So, let's plug in some numbers.

Since we are given that $a_0 = 1$, $a_1 = 213$, and $a_n = 2a_{n-1} + a_{n-2}$, we can recursively calculate a_2 as follows: $a_2 = 2a_1 + a_0 = 2 \cdot 213 + 1 = 427$. Now that we know a_1 and a_2, we can calculate a_3. We have $a_3 = 2a_2 + a_1 = 2 \cdot 427 + 213 = 1067$. Continuing in this way, we see that the first several values of a_n are $\{1, 213,$

427, 1067, 2561, 6189, 14939, 36067, 87073, ...}. Now we can use these a_n values to calculate the partial sum S_8 :

$$S_8 = \sum_{i=1}^{8} \frac{a_{i-1}}{a_i^2 - a_{i-1}^2}$$

$$= \frac{a_0}{a_1^2 - a_0^2} + \frac{a_1}{a_2^2 - a_1^2} + ... + \frac{a_7}{a_8^2 - a_7^2}$$

$$= \frac{1}{213^2 - 1^2} + \frac{213}{427^2 - 213^2} + ... + \frac{36067}{87073^2 - 36067^2}$$

$$\approx 0.00235.$$

If S_8 is approximately 0.00235, then $1/S_8$ is approximately 425.5. Assuming that the infinite sum S converges, it appears that $1/S$ is *approximately* 425. This is just a guess.

One strategy for calculating sums is the *telescoping method* that we learned in Problem 16. Let's try that method with our problem. Each denominator term of S has the form $a_i^2 - a_{i-1}^2$. Since $x^2 - y^2 = (x-y)(x+y)$, a natural approach is to factor the denominator to get $\frac{a_{i-1}}{a_i^2 - a_{i-1}^2} = \frac{a_{i-1}}{(a_i - a_{i-1})(a_i + a_{i-1})}$. In Problem 16, we saw that $\frac{1}{k(k+1)} = \frac{1}{k} - \frac{1}{k+1}$. We need to do something similar for our problem. We need to write something like this:

$$\frac{a_{i-1}}{a_i^2 - a_{i-1}^2} = \frac{a_{i-1}}{(a_i - a_{i-1})(a_i + a_{i-1})} = \frac{A}{a_i - a_{i-1}} - \frac{B}{a_i + a_{i-1}}.$$

This is a classic case of using the *method of partial fractions*. Multiplying through by $(a_i - a_{i-1})(a_i + a_{i-1})$ we get

$$a_{i-1} = A(a_i + a_{i-1}) - B(a_i - a_{i-1})$$
$$= (A - B)a_i + (A + B)a_{i-1}.$$

This means that $A - B = 0$ and $A + B = 1$. So $A = B = 1/2$. We now have

$$\frac{a_{i-1}}{a_i^2 - a_{i-1}^2} = \frac{a_{i-1}}{(a_i - a_{i-1})(a_i + a_{i-1})} = \frac{1}{2(a_i - a_{i-1})} - \frac{1}{2(a_i + a_{i-1})}.$$

This expression still doesn't quite telescope, but we can convert it into the right form by using the given recurrence relation in the form $a_{n+1} = 2a_n + a_{n-1}$, or $a_{n+1} - a_n = a_n + a_{n-1}$. Substituting this into the previous expression, we get

$$\frac{a_{i-1}}{a_i^2 - a_{i-1}^2} = \frac{a_{i-1}}{(a_i - a_{i-1})(a_i + a_{i-1})} = \frac{1}{2(a_i - a_{i-1})} - \frac{1}{2(a_{i+1} - a_i)}.$$

Now the sum S will telescope. That is, most of the terms will cancel each other out. (If you have difficulty seeing this, just plug numbers for a_i into the following expression.)

$$S = \sum_{i=1}^{\infty} \frac{a_{i-1}}{a_i^2 - a_{i-1}^2}$$

$$= \sum_{i=1}^{\infty} \left(\frac{1}{2(a_i - a_{i-1})} - \frac{1}{2(a_{i+1} - a_i)} \right)$$

$$= \frac{1}{2(a_1 - a_0)}$$

$$= \frac{1}{2(213 - 1)}$$

$$= \frac{1}{424}.$$

Since $S = \dfrac{1}{424}$, we have $\dfrac{1}{S} = 424$. $\square$

Our original approximation, using numerical data, suggested that $1/S$ is approximately 425 or 426. But that's because we only used eight terms in the summation for S. In fact, S has infinitely many terms, as stated in the problem.

Problem 18

Logic Days

Which day is two days before the day after the day three days after the day before Tuesday?

This problem is clearly a logic problem, and with some trial and error you can probably figure it out. What makes the problem difficult is the convoluted wording of the problem. Logic problems like this can often be made easier—even trivial—by reformulating them as a mathematics problem.

Basically, we want to convert this word problem into a numerical problem. First, let's assign a number to each day of the week. Let Sunday = 0, Monday = 1, Tuesday = 2, Wednesday = 3, Thursday = 4, Friday = 5, and Saturday = 6. Next, we need to encode the words "before" and "after." If D is the numerical value of a day, such as 2 for Tuesday, then $D-1$ is the day *before* day D and $D+1$ is the day *after* day D. So, this means that the word "before" is a -1 and the word "after" is a $+1$. Now we just need to identify all the occurrences of the words "before" and "after" in the problem statement.

Let D be the required day. We can break the problem statement into pieces by focusing on the "before" and "after" words as follows:

$D =$ (2 days *before*) + (the day *after*) + (the day 3 days *after*) + (the day *before*) + Tuesday.

Now, remembering that Tuesday is day 2, "before" is -1, and "after" is $+1$, we can write the previous equation in numerical form:

$$D = -2 + 1 + 3 - 1 + 2$$
$$= 3.$$

Since 3 = Wednesday, we know that the answer to the logic problem is "Wednesday."

Notice how the logic problem becomes easy once we convert it into a numerical problem. Now that we know this method, we can tackle similar problems of much higher complexity.

Problem 19

Even Powers

If the expression $(x^2 + 2x - 1)^8$ is expanded, what is the sum of the coefficients of the terms with even powers of x?

One way to solve this problem, which should work in theory, is to use the multinomial theorem applied to $(x^2 + 2x - 1)^8$, and then evaluate the resulting multinomial sum, for coefficients of even powers of x, in closed form. Standard mathematical machinery exists for doing this, but that seems like a lot of unnecessary work.

Let's start with a brute-force-dumb solution. Maybe that will give us some insights that will help us find a good solution. Since we have $x^2 + 2x - 1$ taken to the eighth power, and eight is not too big of an exponent, we can simply multiply out the polynomials, and collect like powers of x, to get the following. The calculation takes about one second to perform on a good symbolic calculator like the Texas Instruments TI-89.

$$(x^2 + 2x - 1)^8 = x^{16} + 16x^{15} + 104x^{14} + 336x^{13} + 476x^{12}$$
$$-112x^{11} - 1064x^{10} - 432x^9 + 1222x^8$$
$$+432x^7 - 1064x^6 + 112x^5 + 476x^4$$
$$-336x^3 + 104x^2 - 16x + 1.$$

Now the solution to the problem is simple. We add up the coefficients of the even powers of x, and we get $1 + 104 + 476 - 1064 + 1222 - 1064 + 476 + 104 + 1 = 256$. Right away, you should see a pattern in this answer. This is a classic example of how knowing the answer to a problem often helps us find a solution. Why? The number 256 is a power of 2. More significantly, $256 = 2^8$. If you want to be a good mathematics problem solver, it's good to get familiar with a lot of different number sequences—like powers of two. The fact that the sum of the coefficients of the even powers of x is 256, or 2^8, is very peculiar, because the polynomial $(x^2 + 2x - 1)^8$ has 8 as its exponent.

Now let's look at special cases. For notational convenience, let $f(x) = (x^2 + 2x - 1)^8$. Then $f(1) = 2^8 = 256$ and $f(-1) = (-2)^8 = 256$. Clearly, something interesting is going on with $f(1)$ and $f(-1)$. These observations suggest that there might be a simple, clever solution to this problem.

Let's go back to basics. If we expand $f(x)$, as we did with our brute-force-dumb solution, we get a polynomial. In fact, we get a polynomial of degree 16. We also saw that something

154

strange is going on with $f(1)$ and $f(-1)$ since they are both equal to 256, and 256 is the answer to the problem. So, let's explore a simpler polynomial, say $h(x)$, and see what happens with $h(1)$ and $h(-1)$.

Let $h(x) = ax^4 + bx^3 + cx^2 + dx + e$, for $a \neq 0$. Then $h(1) = a+b+c+d+e$ and $h(-1) = a-b+c-d+e$. Now let's play around with these two expressions. If we add $h(1)$ and $h(-1)$, we get $h(1)+h(-1) = 2a+2c+2e$. Wow! This is just two times the sum of the exponents of the even powers of x. This is just the thing we need to solve our problem. If we divide both sides of the equation by 2, we see that the sum of the coefficients of the even powers of x is

$$a+c+e = \frac{h(1)+h(-1)}{2}.$$

We have discovered the brilliant insight to solve the original problem. The sum of the coefficients of the even powers of x in the expansion of $f(x) = (x^2 + 2x - 1)^8$ is

$$\frac{f(1)+f(-1)}{2} = \frac{(1^2 + 2 \cdot 1 - 1)^8 + ((-1)^2 + 2 \cdot (-1) - 1)^8}{2}$$

$$= \frac{2^8 + (-2)^8}{2}$$

$$= 256.$$

Let's summarize our discoveries as useful theorems:

Theorem 19.1. *Let $f(x)$ be a polynomial in x. The sum of the coefficients of the even powers of x is given by*

$$\frac{f(1) + f(-1)}{2}.$$

Theorem 19.2. *Let $f(x)$ be a polynomial in x. The sum of the coefficients of the odd powers of x is given by*

$$\frac{f(1) - f(-1)}{2}.$$

The process of exploring a problem often leads to key insights that allow us to create a clever solution. Furthermore, during our mathematical exploration, we discovered some results that are significant enough to state as useful theorems. If you happen to discover something interesting during your mathematical problem solving, don't be afraid to state it as a conjecture. Then prove it as a theorem. This is how new mathematics is often discovered.

Problem 20

Coin Flipping

Twenty-five independent, fair coins are tossed in a row. What is the expected number of consecutive head-head pairs?

This is a probability problem involving coin tossing. A fair coin has "heads" (H) on one side and "tails" (T) on the other side. If you toss a *fair* coin, it is equally likely to land with either the head side up or the tail side up. Since we are tossing *independent* coins, the outcome of one coin toss has no effect on the outcome of the other coin tosses. In fact, instead of tossing 25 independent coins, we could just as well toss one fair coin 25 times.

Just to be sure that we understand the problem, let's do an experiment. Let's toss a fair coin 25 times and see what we get. Here is one possible outcome:

H H T H T T T H H H H H T T H H H T H T T H H T H

In this particular experiment, the number of head-head (HH) pairs is 8. We just count, from left to right, the number of times that two heads are next to each other.

The expected number of HH pairs, for 25 coin tosses, is the average number of HH pairs that would occur if we did this experiment millions of times.

This could be a difficult problem to solve, but we will use a strategy that is often useful for these kinds of problems. First, we define a function, $E(n)$, that counts the answer to our problem. Second, we use "divide-and-conquer" to break the outcomes into separate, disjoint cases. Third, we use each of the separate cases to build a recurrence formula for $E(n)$. Finally, we solve for $E(n)$ using whatever method works. In this problem, we will solve for $E(n)$ using our old friend, the *telescoping sum* method. This problem nicely illustrates how mathematical problem solving involves the orchestration of several different ideas and methods in the same problem.

Let $E(n)$ be the expected number of HH pairs for n tosses of a fair coin. We want to find $E(25)$, and that will be the answer to our problem. Right now, we have no idea what $E(n)$ is, but we will build a recurrence relation that expresses $E(n+1)$ in terms of $E(n)$. Recursion is a powerful method for solving problems in mathematics.

We also need to get started somehow. Clearly, if we toss only one coin, then there will be no consecutive HH pairs. So, $E(1) = 0$. Recurrence relations always need two components: A relation or relations that express, say, $E(n+1)$ in terms of $E(n)$

or other terms, and initial conditions, like $E(1) = 0$. You must have some initial conditions for your recurrence relations.

Now we can use a "divide-and-conquer" technique where we break the possible outcomes into separate, disjoint cases. Suppose that we have already tossed the fair coin n times. If we toss the coin one more time, then there are three distinct cases that can occur:

Case 1. On coin toss number $n + 1$, we get a head (H) where the previous coin flip was also an H. If this case occurs, then the number of HH pairs will increase by one. The probability of this case occurring is $1/4$, because the probability of the ordered pair (H, H) occurring is $1/2 \times 1/2 = 1/4$. Remember, for each coin flip, the probability of a head is $1/2$ and the probability of a tail is $1/2$.

Case 2. On coin toss number $n + 1$, we get a head (H) where the previous coin flip was a tail (T). If this case occurs, then the number of HH pairs does not change. This case adds zero new HH pairs. The probability of this case occurring is $1/4$.

Case 3. On coin toss number $n + 1$, we get a tail (T) where the previous coin flip was either H or T. This case adds zero new HH pairs, and it occurs with probability $1/2$. Note that (H, T) has probability $1/4$ and (T, T) has probability $1/4$, so the total probability of Case 3 is $1/4 + 1/4 = 1/2$.

Putting these three cases together, we can build a recurrence relation for $E(n+1)$:

$$E(n+1) = \frac{1}{4}(E(n)+1) + \frac{1}{4}(E(n)+0) + \frac{1}{2}(E(n)+0)$$
$$= E(n) + \frac{1}{4}.$$

This means that, on average, one-quarter of the time the $(n+1)$st coin toss increases the number of HH pairs from $E(n)$ to $E(n)+1$. And three-quarters of the time $(1/4 + 1/2)$ the $(n+1)$st coin toss does not increase the number of HH pairs, so the number of HH pairs stays at $E(n)$, or $E(n)+0$.

   Now we just need to solve the recurrence relation $E(n+1) = E(n) + 1/4$ for $E(n)$. To do this, we write the recurrence relation in a form that will telescope: $E(n+1) - E(n) = 1/4$. Next, we write down a sequence of these recurrence relations as follows:

$$E(n+1) - E(n) = 1/4$$
$$E(n) - E(n-1) = 1/4$$
$$E(n-1) - E(n-2) = 1/4$$
$$\cdots$$
$$E(2) - E(1) = 1/4$$

Now add all of these equations, and notice that lots of terms cancel out. (Remember, $E(1) = 0$.) We end up with the nice result:

$$E(n+1) - E(1) = \frac{1}{4}n,$$
$$E(n+1) - 0 = \frac{1}{4}n,$$
$$E(n+1) = \frac{1}{4}n.$$

It is now easy to solve our problem. We have $E(25) = \frac{1}{4} \cdot 24 = 6$.

This is the answer to our problem. If 25 independent fair coins are tossed in a row, the *expected* number of consecutive HH pairs is 6. □

Another way to think of this result, in practical terms, is that if we flip a fair coin 25 times and count the number of HH pairs, then repeat the experiment, say, 1,000,000,000 times, *on average* we will have 6 HH pairs for the experiment of 25 coin tosses.

Problem 21

Equal Sums

Each cell in a 3 x 3 arrangement of cells is randomly filled with one of the numbers −1, 0, or 1. Prove that of the eight possible sums along the rows, columns, and diagonals, at least two sums must be equal.

Let's draw a picture:

-1	0	1
1	1	-1
-1	0	0

This picture shows one possible way to assign numbers −1, 0, or 1 to cells. Of course, many other assignments are possible. There are seven possible sums that we could get for a row, column, or diagonal. We could get a sum of -3, -2, -1, 0, 1, 2, or 3. For example, in the picture above the sum of the numbers in the first row is $-1+0+1=0$. There are also 3 rows, 3 columns, and 2 diagonals, which we could call "line sums" for convenience.

We want to prove that no matter how we assign the numbers −1, 0, or 1 to the nine cells there will always be at least two line sums that are equal. A brute-force-dumb way to do this would be to simply construct all possible configurations, and then show that for each possible configuration there are at least two equal line sums. The trouble with this approach is that there are $3^9 = 19683$ possible 3 x 3 configurations. That's way too many, and we really need a clever way to solve this problem.

The key to solving this problem, and similar problems, is an important mathematical technique known as the *pigeonhole principle* (also known as the *Dirichlet box principle*). The pigeonhole principle is deceptively simple. In its simplest form, the pigeonhole principle states that if you put three pigeons into two pigeonholes, then at least one hole will contain at least two pigeons.

Let's generalize the pigeonhole principle so that we can use it to solve many difficult mathematics problems.

Theorem 21.1 (Pigeonhole Principle). *If $nk + 1$ pigeons are put into n pigeonholes, then at least one hole will contain at least $k + 1$ pigeons.*

The proof of this theorem is quite easy if we use the method of *proof by contradiction*. Suppose the theorem is false. Then if we place $nk+1$ pigeons into n pigeonholes, no pigeonhole will contain $k+1$ or more pigeons. This means that each of the n pigeonholes will have *at most k* pigeons. The total number of pigeons is then *at most nk*, but this contradicts the fact that we have $nk+1$ pigeons.

Let's use the pigeonhole principle to solve our problem. The usual difficulty in applying the pigeonhole principle is deciding what things are the "pigeons" and what things are the "holes." Clearly, the "pigeons" and "holes" are metaphors for almost any kind of mathematical objects or structures.

What mathematical structures are relevant to our problem? As discussed previously, we have 7 possible numerical sums (-3, -2, -1, 0, 1, 2, or 3) and 8 possible lines (3 rows + 3 columns + 2 diagonals). Since each of the 8 lines will have one of the 7 possible line sums, we can use the following assignment:

- Pigeons = lines (rows, columns, or diagonals)
- Pigeonholes = line sums (-3, -2, -1, 0, 1, 2, 3)

So, for our problem, we have 8 pigeons being placed into 7 pigeonholes. By Theorem 21.1, this means that $nk+1=8$ and $n=7$, so $k=1$. Then Theorem 21.1 tells us that *at least* one pigeonhole contains *at least* $k+1=2$ pigeons. In other words, at least one line sum (-3, -2, -1, 0, 1, 2, or 3) occurs in at least two lines (rows, columns, or diagonals). Using the pigeonhole principle, we have proved that of the eight possible sums along the rows, columns, and diagonals, at least two sums must be equal. This is true regardless of how we assign the numbers −1, 0, or 1 to the cells. □

Anytime a problem asks you to *prove* that *at least* some number of something occurs, for some kind of discrete mathematical configuration, that should be a "red flag" hinting that you should try using the pigeonhole principle. You should memorize Theorem 21.1.

Problem 22

Divisible by Five

Prove that the number $N = 1 + 2^{99} + 3^{99} + 4^{99} + 5^{99}$ is divisible by 5.

A number is divisible by 5 if its last digit is either 0 or 5. So, one way to solve this problem is to calculate N and look at its last digit. This approach will work, in theory, but it is not practical. Since we would need to calculate numbers taken to the 99th power, N is going to have about 70 digits. That's a really big number. We need a better a approach to solve this problem.

In the interest of symmetry, which is generally a good thing in mathematics, let's write the number 1 as 1^{99}. Then every term in N will be a power of 99, and we have $N = 1^{99} + 2^{99} + 3^{99} + 4^{99} + 5^{99}$. It is generally a good idea in mathematical problem solving to try to represent things in the most symmetrical way.

Now, the exponents "99" are too big to help us with our investigation. So let's first generalize the number N as a function of some positive integer m. We can write this as follows:

$$N(m) = 1^m + 2^m + 3^m + 4^m + 5^m.$$

One dirty little secret to solving mathematics problems is that it is often better to consider a more general problem than the one you are actually trying to solve. Here, we have generalized the problem to a consideration of $N(m)$. Now we can look at small cases and special cases to see if there are any patterns that we can exploit.

First, consider the special case of $m = 1$. We then have $N(1) = 1 + 2 + 3 + 4 + 5 = 15$, which is clearly divisible by 5. We can also rearrange the terms and group the summands as follows:

$$N(1) = (1 + 4) + (2 + 3) + 5.$$

Now we can easily see that $N(1)$ is divisible by 5 because $1 + 4$, $2 + 3$, and 5 are all divisible by 5. This rearrangement of the summands suggests a bold idea. If we write $N(m)$ as $(1^m + 4^m)$ $+ (2^m + 3^m) + 5^m$, then maybe $1 + 4$ divides $(1^m + 4^m)$ and $2 + 3$ divides $2^m + 3^m$. And, of course, 5 divides 5^m. This is wishful thinking, but it might work. More generally, is it true that $a + b$ divides $a^m + b^m$ for positive integers a, b, and m? Let's investigate this idea and see if it works.

Again, let's look at small cases and special cases. Does $2 + 3$ divide $2^1 + 3^1$? Yes. Does $2 + 3$ divide $2^2 + 3^2 = 13$? No. Does $2 + 3$ divide $2^3 + 3^3 = 35$? Yes. You can continue with this investigation, and you will discover that $2 + 3$ appears to divide

$2^m + 3^m$ whenever m is odd. This observation gives us the following conjecture:

Conjecture: For positive integers a, b, and m, if m is odd then $a + b$ divides $a^m + b^m$.

Now our challenge is to prove this conjecture. There may be many ways to prove this, so there is no single, right way. We could, for example, try to prove the conjecture using our old friend, mathematical induction. Or, we could look for an existing theorem that might help. Here is one useful theorem called the "Factor Theorem":

Theorem 22.1 (Factor Theorem). *If k is a zero of the polynomial $f(x)$, then $x - k$ is a factor of $f(x)$.*

To use this theorem, let $f(x) = a^n + x^n$. This is certainly a polynomial. By the Factor Theorem, if $-a$ is a zero of $f(x)$, which means that $f(-a) = 0$, then $x - (-a) = x + a$ is a factor of $f(x)$. Note that $f(-a) = a^n + (-a)^n$, and $a^n + (-a)^n = a^n - a^n = 0$ if n is odd. This means that $x + a$ divides $x^n + a^n$. Now let $x = b$ and we immediately have the result we need to solve our problem: if n is odd, then $a + b$ is a factor, or divisor, of $a^n + b^n$.

What if n is even? If n is even, then *sometimes* $a+b$ divides $a^n + b^n$. For example, for $a = b = 2$ and $n = 2$, we see that $2 + 2$ certainly divides $2^2 + 2^2$.

In our problem, $N(99) = 1^{99} + 2^{99} + 3^{99} + 4^{99} + 5^{99} = (1^{99} + 4^{99}) + (2^{99} + 3^{99}) + 5^{99}$. Since 99 is odd, we know that $1 + 4$ divides $(1^{99} + 4^{99})$ and $2 + 3$ divides $(2^{99} + 3^{99})$. This means that 5 divides $(1^{99} + 4^{99})$, $(2^{99} + 3^{99})$, and 5^{99}. Therefore, 5 divides N. That solves the original problem. $\square$

Problem 23

A Diophantine Equation

Solve the equation $\dfrac{1}{m}+\dfrac{1}{n}+\dfrac{3}{mn}=\dfrac{1}{4}$ for positive integers m and n.

An equation that we want to solve in the positive integers is called a *Diophantine equation*. Generally, these kinds of equations are very difficult to solve. Let's start by letting $f(m, n) = \dfrac{1}{m}+\dfrac{1}{n}+\dfrac{3}{mn}$. Then our problem is to find positive integers m and n, if they exist, for $f(m, n) = 1/4$.

The first thing to notice is that $f(m, n)$ is a *symmetric* function. In other words, $f(m, n) = f(n, m)$. So, if we can find a solution (m, n), then we will also have the solution (n, m).

Let's begin by looking at a special case. Is there a solution to $f(m, n) = 1/4$ if $m = n$? If $m = n$, the equation $f(m, n) = 1/4$ becomes $\dfrac{2}{n}+\dfrac{3}{n^2}=\dfrac{1}{4}$. After a little algebraic manipulation we get $n^2 - 8n - 12 = 0$. We can now find n using the quadratic formula.

Theorem 23.1 (Quadratic Formula). *Let* $ax^2 + bx + c = 0$, *for real numbers a, b, and c, and* $a \neq 0$. *Then the roots of the equation are given by* $x = \dfrac{-b \pm \sqrt{b^2 - 4ac}}{2a}$.

Using the quadratic formula for $n^2 - 8n - 12 = 0$, we find that if $m = n$ then $n = 4 \pm 2\sqrt{7}$. Although this is a solution to the equation $n^2 - 8n - 12 = 0$, we can't use this solution because $4 \pm 2\sqrt{7}$ are not positive integers. Remember, we want to find *positive integers* (m, n) such that $f(m, n) = 1/4$. We conclude, therefore, that m cannot equal n. This leaves two possibilities: either $m > n$ or $m < n$ (if a solution exists). Not all Diophantine equations have solutions in the positive integers. When we attempt to solve a Diophantine equation, several kinds of questions naturally arise: Does a solution exist? If a solution exists, is it unique? If solutions exist, how can we find them?

What else can we discover by looking at the equation $f(m, n) = 1/4$? Don't be afraid to make simple observations. Sometimes simple observations are the key to solving your problem. When we look at the equation $\dfrac{1}{m} + \dfrac{1}{n} + \dfrac{3}{mn} = \dfrac{1}{4}$, one thing that may not be entirely obvious is that m and n must each be greater than 4. Why? If either m or n is equal to 4 (or less), then the left-hand side of the equation $\dfrac{1}{m} + \dfrac{1}{n} + \dfrac{3}{mn} = \dfrac{1}{4}$ will be greater than $1/4$, while the right-hand side of the equation is equal to $1/4$. That would be a contradiction. So, we must have both $m > 4$ and $n > 4$.

Since $m > 4$ and $n > 4$, we can add some unknown positive integers, say a and b, to 4 to get m and n. In other words, there exist positive integers a and b such that $m = a + 4$ and $n = b + 4$. These positive integers, a and b, are called *slack variables.* It's another sneaky tool to add to your problem solving bag of tricks. To convert an inequality, like $m > 4$, into an equality, we add a slack variable. Then we get $m = a + 4$.

With $m = a + 4$ and $n = b + 4$, the equation $f(m, n) = 1/4$ becomes $\dfrac{1}{(a+4)} + \dfrac{1}{(b+4)} + \dfrac{3}{(a+4)(b+4)} = \dfrac{1}{4}$. If we now multiply both sides of the equation by $4(a+4)(b+4)$ and simplify, we get the amazingly simple result that $ab = 28$. If we can solve the equation $ab = 28$ for positive integers a and b, then we can find m and n because $m = a + 4$ and $n = b + 4$.

The number 28 can be factored into a product of two positive integers in several ways: 1×28, 28×1, 2×14, 14×2, 4×7, and 7×4. This gives us six solutions for (a, b): $(1, 28)$, $(28, 1)$, $(2, 14)$, $(14, 2)$, $(4, 7)$ and $(7, 4)$. Then, since $m = a + 4$ and $n = b + 4$, we get six solutions to the original equation $f(m, n) = 1/4$. The six solutions to the original problem are $(5, 32)$, $(32, 5)$, $(6, 18)$, $(18, 6)$, $(8, 11)$, and $(11, 8)$. □

We should check that each of these solutions actually satisfy the equation $f(m, n) = 1/4$. For example, for the ordered pair $(m, n) = (5, 32)$, we have

$$
\begin{aligned}
f(5, 32) &= \frac{1}{5} + \frac{1}{32} + \frac{3}{5 \cdot 32} \\
&= \frac{1}{5} + \frac{1}{32} + \frac{3}{160} \\
&= \frac{1}{4}.
\end{aligned}
$$

Problem 24

A Functional Equation

Solve the equation $x f(x) + 2x f(-x) = -1$ for $f(x)$.

This kind of problem is called a *functional equation*. We are given an equation that involves some unknown function $f(x)$, and the problem is to find the function $f(x)$ that satisfies the equation. We are assuming, of course, that the function $f(x)$ exists. It is also possible that there may be more than one function $f(x)$ that satisfies the given equation.

Functional equations are generally difficult to solve. There is still a great deal of research being done in this area, and a unified theory does not yet exist. Unfortunately, there is no known universal method for solving functional equations. Unless a functional equation has a special, well-studied form, our approach to solving these equations involves a collection of *ad hoc* tricks and lucky guesses. Some of the methods used include substitution, elimination, undetermined coefficients, and recursion. We can also solve some functional equations by looking at special values.

Here are a few of the many techniques for solving functional equations:

1. Look at special values like $f(0)$, $f(1)$, and $f(-1)$.

2. Consider $f(x+k)$ and $f(kx)$ for some constant k.

3. Replace x by $-x$ and see what happens.

4. Replace x by some function of x like $1/x$.

5. See if the function is *multiplicative*, which means that $f(mn) = f(m)f(n)$ whenever m and n are coprime positive integers.

This list is by no means comprehensive. It just gives us some ideas on how to explore a given functional equation. If we are lucky, special transformations will reduce the given functional equation to a system of linear algebraic equations. In that case, we can use standard methods from linear algebra to solve for $f(x)$.

As always, let's first look at some special cases. If $x = 1$, we get $f(1) + 2f(-1) = -1$. Well, I'm not sure what this tells us. So, let's try another special case. If $x = 0$, the functional equation $x f(x) + 2x f(-x) = -1$ becomes $0 \cdot f(0) + 0 \cdot f(-0) = -1$, or simply $0 = -1$. This, of course, is impossible. But this actually tells us something useful. Since x cannot equal 0, because it results in a contradiction, it is likely that the unknown function $f(x)$ has a "divide-by-zero" condition. In other words, maybe

$f(x)$ has x in the denominator, like $f(x) = \dfrac{ax+b}{x}$, or something like that. Since division by zero is undefined, this could explain why x can't be zero.

What else can we do? Let's try replacing x by $-x$ and see what happens. If we replace x by $-x$, the equation $x\,f(x) + 2x\,f(-x) = -1$ becomes $-x\,f(-x) - 2x\,f(x) = -1$. This might look confusing, but we are actually making progress. We have two equations and two unknowns. The two "unknowns" are $f(x)$ and $f(-x)$. This becomes clear if we replace $f(x)$ by u and $f(-x)$ by v. Then we get the following system of simultaneous linear equations:

$$\begin{cases} xu + 2xv = -1 \\ -2xu - xv = -1 \end{cases}$$

In matrix form, the system of equations looks like this:

$$\begin{bmatrix} x & 2x \\ -2x & -x \end{bmatrix} \begin{bmatrix} u \\ v \end{bmatrix} = \begin{bmatrix} -1 \\ -1 \end{bmatrix}$$

At this point, the problem is essentially solved. We need to find u and v, where $u = f(x)$ and $v = f(-x)$. There are several ways to do this. You can solve the first equation for u in terms of v, substitute for u in the second equation, and solve for v. Another approach is to find the multiplicative inverse of the matrix

$$\begin{bmatrix} x & 2x \\ -2x & -x \end{bmatrix}, \text{ which is } \begin{bmatrix} \dfrac{-1}{3x} & \dfrac{-2}{3x} \\ \dfrac{2}{3x} & \dfrac{1}{3x} \end{bmatrix}, \text{ then left-multiply the matrix}$$

equation by the inverse matrix to obtain

$$\begin{bmatrix} u \\ v \end{bmatrix} = \begin{bmatrix} \dfrac{-1}{3x} & \dfrac{-2}{3x} \\ \dfrac{2}{3x} & \dfrac{1}{3x} \end{bmatrix} \begin{bmatrix} -1 \\ -1 \end{bmatrix} = \begin{bmatrix} \dfrac{1}{x} \\ \dfrac{-1}{x} \end{bmatrix}$$

This is just standard linear algebra. The hard part was figuring out how to get the functional equation into a system of simultaneous linear equations, and we did that with the lucky transformation of replacing x by $-x$.

It is now clear that $u = f(x) = \dfrac{1}{x}$. That's the solution to the problem. We can also see why x cannot be zero, since division by zero is undefined. □

Problem 25

Exponential Equation

Solve the equation $\left|x-3\right|^{\frac{(x^2-8x+15)}{(x-2)}} = 1$.

Sometimes a problem is simpler than it looks. Here, we are taking the quantity $\left|x-3\right|$ to the power $(x^2-8x+15)/(x-2)$. This looks like a difficult problem, so let's go back to basics.

Instead of considering the nasty functions in the problem statement, consider the simpler problem $a^b = 1$ for some real numbers a and b. How can $a^b = 1$? Just play around with it. First, note that 0^0 is undefined. So we cannot have $a = b = 0$. Next, note that if $a^b = 1$, then either $a = 1$ or $b = 0$. A nonzero number taken to the 0 power is 1. For example, $3^0 = 1$.

Since either $a = 1$ or $b = 0$, let's start by setting $a = \left|x-3\right| = 1$. This means that either $x - 3 = 1$ or $x - 3 = -1$. This implies that either $x = 4$ or $x = 2$. Now we have to think a bit about each of these possible x values. Unfortunately, we must discard $x = 2$, because if we plug $x = 2$ into the original equation, we find that $(x^2 - 8x + 15)/(x-2)$ will result in division by zero. Division by zero is not defined in mathematics. So $x = 4$ is one solution. We can easily verify this by substituting $x = 4$ into the original equation:

$$|4-3|^{\frac{(4^2-8\cdot4+15)}{(4-2)}} = 1^{\frac{-1}{2}} = 1.$$

Next, we consider the possibility of $b = 0$. We set
$b = \dfrac{(x^2 - 8x + 15)}{(x - 2)} = 0$, or $x^2 - 8x + 15 = 0$. Factoring this

polynomial, we get $x^2 - 8x + 15 = (x-3)(x-5) = 0$. The roots are $x = 3$ and $x = 5$. Each of these is a candidate solution to the original equation, but we need to check each of them. When we do this, we see that $x = 3$ is not a valid solution because, when we substitute $x = 3$ into the equation $|x-3|^{\frac{(x^2-8x+15)}{(x-2)}} = 1$, we get 0^0, which is undefined. However, $x = 5$ is a valid solution:

$$|5-3|^{\frac{(5^2-8\cdot5+15)}{(5-2)}} = 2^0 = 1.$$

In conclusion, the solutions to the equation $|x-3|^{\frac{(x^2-8x+15)}{(x-2)}} = 1$ are $x = 4$ and $x = 5$. □

Problem 26

Absolute Value

Let x, y, and z be real numbers that satisfy the following system of nonlinear equations:

$$x^2 + 6y = -14$$
$$y^2 + 12z = -63$$
$$z^2 - 4x = 28$$

Find the value of $|x + y + z|$.

We want to find the absolute value of $x + y + z$. We could try solving each of the equations for x, y, and z, but then we would still need to know x^2, y^2, and z^2. That doesn't help much.

Let's look at the definition of the absolute value function. One definition states that $|x| = x$ if $x \geq 0$ and $|x| = -x$ if $x < 0$. Another, equivalent, definition states that $|x| = \sqrt{x^2}$. Let's see what happens if we use this definition. We get $|x + y + z| = \sqrt{(x+y+z)^2}$, and $(x+y+z)^2 = x^2 + y^2 + z^2 + 2xy + 2xz + 2yz$. This last expression contains x^2, y^2, and z^2, just like our problem does. So maybe we should *add* the given equations to obtain $x^2 + y^2 + z^2$. If we add the given equations, we obtain one equation:

$$x^2 + 6y + y^2 + 12z + z^2 - 4x = -49.$$

It's a good habit to collect like variables together. Doing so, we get the following:

$$(x^2 - 4x) + (y^2 + 6y) + (z^2 + 12z) = -49.$$

Now what can we do? One idea is to factor each of the polynomials, but how? If we had, say, $x^2 - 4x + 4$, instead of $x^2 - 4x$, then we could factor it as $x^2 - 4x + 4 = (x-2)^2$. In order for this to work, we need to add and subtract 4 to $x^2 - 4x$. Similarly, we can add and subtract numbers to each of $y^2 + 6y$ and $z^2 + 12z$ so that they will factor:

$$(x^2 - 4x + 4) - 4 + (y^2 + 6y + 9) - 9 + (z^2 + 12z + 36) - 36 = -49.$$

Rearranging this expression, we get

$$(x^2 - 4x + 4) + (y^2 + 6y + 9) + (z^2 + 12z + 36) = 0.$$

Now we can factor each of the polynomials to obtain

$$(x-2)^2 + (y+3)^2 + (z+6)^2 = 0.$$

We still don't know the values of x, y, and z, and it may appear that we are stuck. Remember, however, that *no square is negative.* In other words, $(x-2)^2 \geq 0$, $(y+3)^2 \geq 0$, and $(z+6)^2 \geq 0$. The only way that the sum of all these squares can equal zero is if each of them equals zero: $(x-2)^2 = 0$, $(y+3)^2 = 0$, and $(z+6)^2 = 0$. Now the solution is trivial. We must have $x = 2$, $y = -3$, and $z = -6$. Finally, we note that $|x+y+z| = |2-3-6| = |-7| = 7$. This is the final answer to the problem. □

Problem 27

Find the Exponents

Solve the equation $9^x - 3^{x+1} - 4 = 0$, for real x.

Ordinarily, to solve for exponents in a problem we need to use logarithms. For example, if $y = 9^x$, then $\log y = \log(9^x) = x\log 9$. Then we can solve for x to get $x = \log y / \log 9$. It doesn't matter what the base of the logarithm is. We will get the same answer for any base of logarithm. The difficulty with our problem is that we can't take the logarithm of the given equation directly. The statement $\log(9^x - 3^{x+1} - 4) = \log 0$ doesn't make sense, because the logarithm of zero is undefined. Also, we cannot distribute the logarithm among the terms in $9^x - 3^{x+1} - 4$. Logarithms are not distributive over addition. In other words, $\log(9^x - 3^{x+1} - 4) \neq \log 9^x - \log 3^{x+1} - \log 4$. So, how do we proceed?

First, notice a small detail. The given equation has both a 3 and a 9. And $9 = 3^2$. It is often the case that the numbers that appear in mathematical problems are not random. They are often related to each other in some special way so that the mathematics works out perfectly. If you can discover how the numbers are related to each other, that may help you solve the problem. Don't be afraid to make simple, obvious observations. Here, since $9 = 3^2$, we can rewrite the original equation as $3^{2x} - 3^{x+1} - 4 = 0$, or $(3^x)^2 - 3(3^x) - 4 = 0$. Now we can use substitution. We introduce a new variable, say u, and let $u = 3^x$. Then the equation can be written in terms of the new variable u as $u^2 - 3u - 4$. This is a polynomial in the variable u, and we can factor it as $u^2 - 3u - 4 = (u+1)(u-4) = 0$. The only way that the product $(u+1)(u-4)$ can equal zero is if either $u+1=0$ or $u-4=0$ (or both). This means that either $u = -1$ or $u = 4$.

Remember, we defined $u = 3^x$, and we know that either $u = -1$ or $u = 4$. That gives us two equations: $-1 = 3^x$ or $4 = 3^x$. Now we can take logarithms to solve for x. We have (1) $\log_3(-1) = x$; and (2) $\log_3(4) = x$. Since $\log_3(-1)$ is imaginary (not real), the only solution to our problem is $x = \log_3 4 \approx 1.26186$. $\square$

You can verify on a calculator that $9^{1.26186} - 3^{2.26186} - 4 = 0.000010829$. The exact answer is $\log_3 4 = \ln 4 / \ln 3$.

We used logarithms to the base 3 because $\log_3 3 = 1$. But you could have used a logarithm to any base and things would still work out just fine. That's because one of the rules of logarithms states that $\log_3 a = \dfrac{\log_b a}{\log_b 3}$ for any base $b > 1$.

Problem 28

Congruent Angles

Given rectangle *ABCD*, as shown, points *M* and *N* are the midpoints of *BC* and *CD,* respectively. Point *P* is at the intersection of line segments *BN* and *DM*. Prove that angle *MAN* is congruent (equal) to angle *BPM*.

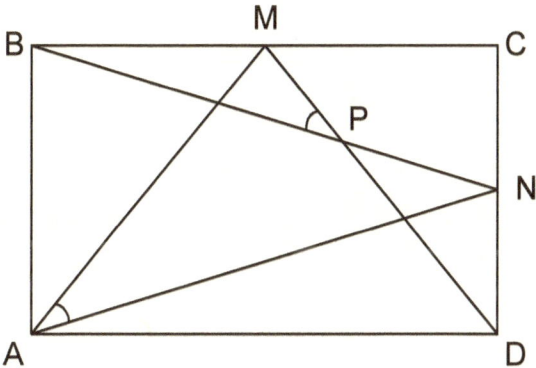

Geometry problems are both interesting and challenging. They can be quite difficult to solve. The brute-force-dumb way to solve a geometry problem is to start labeling every edge and angle that you know, and then start calculating everything that you don't know. The calculations can be done using the Pythagorean Theorem, the Law of Sines, and the Law of Cosines (see Appendix B). Hopefully, after you calculate everything that you possibly can, at some point the quantity that you want will appear. This rather brutish and ugly technique is sometimes called *angle chasing*.

If there is a secret to solving geometry problems, it is the technique of adding the right *auxiliary construction* to your geometric figure. An auxiliary construction is an additional point, line segment, line, ray, or circle that you add to the figure to reveal a hidden aspect of the problem that is necessary to solve it.

In this problem, we add the line segment QD to the figure, as shown by the dotted line. QD is constructed such that it is parallel to line segment BN. So QD and BN are parallel. Now, MD is a transversal that cuts the parallel lines BN and QD. Therefore, corresponding angles are congruent, and angle BPM is congruent to angle QDM. By symmetry, angle MAN is congruent to angle QDM, which is congruent to BPM. Therefore, angle MAN is congruent to angle BPM. We write this as $\angle MAN \cong \angle BPM$. □

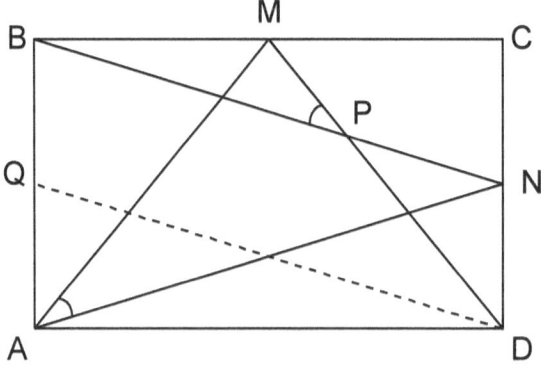

Problem 29

Angle Bisection

Point *P* is at the center of a square constructed on the hypotenuse *AC* of the right-angled triangle *ABC*. Prove that the line segment *BP* bisects angle *ABC*.

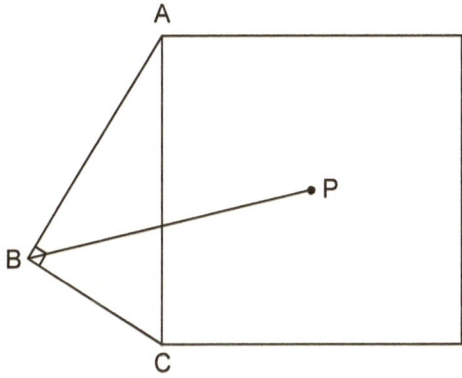

Most geometry problems are carefully constructed so that some key piece of information is hidden from view. To solve the problem, you need to find the hidden information by creating some *auxiliary construction*. A good way to do this is to try to create some symmetry in the problem. I wish there was an easy way to always do this, but it basically takes some practice. For this problem, the insightful move is to construct a circle about the center of segment *AC*.

Angle *ABC* is given as 90 degrees, and you may remember from geometry that the angle in a semi-circle is a right angle. We can construct a semi-circle about the bisector, point *O*, of segment *AC*. That semi-circle will contain points *A*, *B*, and *C*. It's also true that for any square, if we connect the corners to the center (*A* to *P* and *C* to *P*), then we will get a right angle at *P*. So, angle *APC* is also 90 degrees.

Consider the figure below, where we have added a circle as an auxiliary construction. Since angle *ABC* is 90 degrees (given) and angle *APC* is 90 degrees, we can draw a circle about point *O* through points *B* and *P*. Since arc *AP* equals arc *PC*, the angles subtended by these arcs are equal. Therefore, $\alpha = \beta$. Since $\alpha = \beta$, line segment *BP* bisects angle *ABC*. □

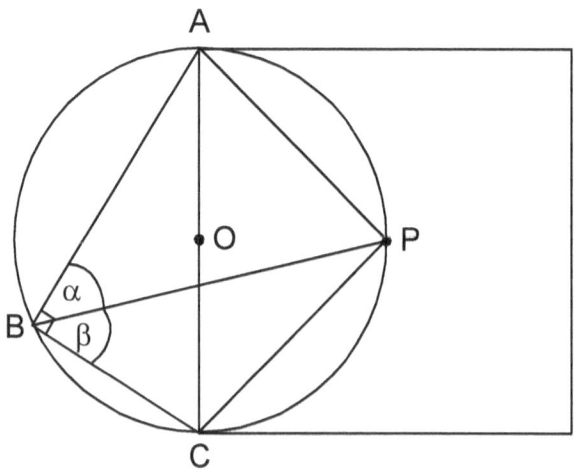

Two key geometry theorems are relevant to this problem:

Theorem 29.1 (Thales' Theorem). *An angle in a semi-circle, constructed on the diameter of the circle, is a right angle. (Examples: Angles ABC and APC are angles constructed on the circle diameter AC. Both ABC and APC are right angles.)*

Theorem 29.2. *Angles subtending congruent arcs of a circle are congruent. (Example: Since arc AP is congruent to arc PC, angle α is congruent to angle β.)*

Problem 30

Order by Averaging

What is the nth term of the infinite sequence $-4, 7, -4, 7, \ldots$?

The given sequence consists of alternating terms -4 and 7. Presumably, the nth term can be represented by some function of n, say $f(n)$, and our task is to find a formula for the function $f(n)$. Looking at the give sequence, we see that $f(1) = -4$, $f(2) = 7$, $f(3) = -4$, $f(4) = 7$, and so on. There is an obvious pattern here: $f(odd) = -4$ and $f(even) = 7$.

There are some standard mathematical tricks for handling problems like this, but I would like to demonstrate how the technique of *averaging* often imposes additional order on a problem. Sometimes the average value of a variable is better behaved than the original variable.

For this problem, the average value of any two consecutive terms in the sequence is $a = (-4+7)/2 = 3/2$. Next, note that $-4 < 3/2 < 7$. Also, the average $a = 3/2$ lies exactly in the middle, in between, -4 and 7, since $3/2 - (-4) = 11/2$ and $7 - 3/2 = 11/2$. These are nice observations, but what can we do with them?

The key idea is to *subtract* $3/2$ (the average) from each term in the sequence $-4, 7, -4, 7, ...$ If we do this, we get the modified sequence $-5.5, 5.5, -5.5, 5.5, ...$ This new sequence consists of terms that alternate between -5.5 and 5.5. This new sequence is easier to understand than the original sequence because every term in the sequence is ± 5.5. We just need to figure out how to generate alternating plus and minus signs. That's easy. We simply note that $(-1)^n = -1$ for odd integers n, and $(-1)^n = +1$ for even integers n. This means that the nth term in the *new sequence* $-5.5, 5.5, -5.5, 5.5, ...$ is $h(n) = (-1)^n \cdot 5.5$. So, we have figured out a formula for the nth term of the new, modified sequence. But how can we find a formula for $f(n)$, which is the nth term of the original sequence? Let's think for a minute. How did we form the new sequence? We formed the new sequence by *subtracting* $3/2$ from each term in the sequence. To get the original sequence back again, we just *add* $3/2$ to each term of the modified sequence $-5.5, 5.5, ...$ This means that the desired formula for $f(n)$ is just

$$f(n) = (-1)^n \cdot 5.5 + 3/2.$$

We can easily test this formula. For example, for $n = 1$, we have $f(1) = (-1)^1 \cdot 5.5 + 1.5 = -4$, and $f(2) = (-1)^2 \cdot 5.5 + 1.5 = 7$, and so on.

Problem 31

A Trigonometric Identity

Prove the following trigonometric identity:

$$\cos\frac{\theta}{2}\cos\frac{\theta}{4}\cos\frac{\theta}{8}...\cos\frac{\theta}{2^n} = \frac{\sin\theta}{2^n \sin\dfrac{\theta}{2^n}}.$$

When I see a problem like this, it's time to pour a glass of vodka to calm my nerves. It brings back nightmares from high school trigonometry class. How can we possibly solve a problem like this? We are asked to prove something. Most likely, we will need to find some suitable standard trigonometric identity, like the ones in the Trigonometry section of Appendix B, and do some clever manipulation like adding, multiplying, or using back-substitution. That's usually how these things work. But which elementary trigonometry identity should we use?

By now, you probably get the idea of how to explore a mathematics problem. We start by looking at small cases, special cases, and extreme cases. A good general strategy is to *let the problem suggest its own solution!* Instead of getting dazzled by the general parameter n, what happens when $n = 1$?

When $n = 1$, the given identity becomes $\cos\dfrac{\theta}{2} = \dfrac{\sin\theta}{2\sin\dfrac{\theta}{2}}$, or

simply $2\sin\dfrac{\theta}{2}\cos\dfrac{\theta}{2} = \sin\theta$. Before going further, let's simplify

this last expression by substituting $\varphi = \dfrac{\theta}{2}$. We then get

$2\sin\varphi\cos\varphi = \sin(2\varphi) = \sin(\varphi + \varphi)$. Aha! That's it. This is

indeed a simple, fundamental trigonometric identity that you

(should) already know. So we know that the case for $n = 1$ is

certainly true. There is no escaping the need to remember some

important theorems, like the ones in Appendix B. I wish I could

say that you don't need to memorize anything at all, but, if that

were the case, then you wouldn't know anything at all either.

You just can't win.

The problem has told us what trigonometric identity we need

to use to solve the problem: $\cos\dfrac{\theta}{2} = \dfrac{\sin\theta}{2\sin\dfrac{\theta}{2}}$. What do we do

now? It's rather straightforward from here. With this identity,

we replace θ by $\theta/2$ to get $\cos\dfrac{\theta}{4} = \dfrac{\sin\dfrac{\theta}{2}}{2\sin\dfrac{\theta}{2^2}}$. Now we do it

again. Replacing θ by $\theta/2$ again, we get $\cos\dfrac{\theta}{8} = \dfrac{\sin\dfrac{\theta}{2^2}}{2\sin\dfrac{\theta}{2^3}}$. We

can clearly continue with this process of successively replacing

θ by $\theta/2$. Now we just multiply all these little identities together to get

$$\cos\frac{\theta}{2}\cos\frac{\theta}{4}\cos\frac{\theta}{8}...\cos\frac{\theta}{2^n} = \frac{\sin\theta}{2\sin\dfrac{\theta}{2}} \cdot \frac{\sin\dfrac{\theta}{2}}{2\sin\dfrac{\theta}{2^2}} \cdot \frac{\sin\dfrac{\theta}{2^2}}{2\sin\dfrac{\theta}{2^3}} ... \frac{\sin\dfrac{\theta}{2^{n-1}}}{2\sin\dfrac{\theta}{2^n}}$$

$$= \frac{\sin\theta}{2^n \sin\dfrac{\theta}{2^n}} \cdot \square$$

Notice that a lot of things cancel out to give us a nice, simple answer. This often happens in mathematics. God is a mathematician. The journey is long and arduous. The suffering is great. But the final results are simple and beautiful.

To solve a mathematics problem, you must *explore the problem.* A good place to begin exploring your problem is to first examine small cases, special cases, and extreme cases.

Problem 32

An Even Product

Let $a_1, a_2, \ldots, a_n$ be an arbitrary rearrangement of the numbers 1, 2, $\ldots$, n. Prove that if n is odd, then the product $(a_1 - 1)(a_2 - 2)(a_3 - 3)\ldots(a_n - n)$ is even.

In this problem, the sequence $a_1, a_2, \ldots, a_n$ can be *any* rearrangement of the numbers 1, 2, $\ldots$, n. For example, for $n = 5$, the sequence could be 3, 5, 1, 4, 2. Or it could be 1, 2, 3, 4, 5. It could be any permutation of the numbers 1 through 5. So we are being asked to prove something that is a mathematical *invariant*—a quantity or property that does not change as the problem changes.

The problems wants us to show that *if n is odd*, then a certain product is *even*. The problem is almost shouting at us to consider *parity*—even and odd properties of numbers. Let's start by reviewing some basing facts about parity of positive integers. We will just state these facts, but they are easy to prove or verify.

- Property 1: An even number plus an even number is an even number. Example: $2 + 4 = 6$.

- Property 2: An odd number plus an odd number is an even number. Example: $3 + 5 = 8$.

- Property 3: An odd number plus an even number is an odd number. Example: $3 + 6 = 9$.

- Property 4: The sum of an odd number of odd numbers is an odd number. Example: $1 + 7 + 3 = 11$.

- Property 5: The product of two even numbers is even. Example: $2 \times 4 = 8$.

- Property 6: The product of two odd numbers is odd. Example: $3 \times 5 = 15$.

- Property 7: The product of an even number and an odd number is even. Example: $4 \times 5 = 20$.

Now let's return to our problem. The product that we want to evaluate contains the terms $a_1 - 1$, $a_2 - 2$, ... , $a_n - n$. We can assume that n is odd since that is what the problem wants us to consider. There are only two basic things that we can do with the terms $a_k - k$. We can add them or we can multiply them. So, let's start with the first possibility. What happens if we add the terms $a_1 - 1$, $a_2 - 2$, ... , $a_n - n$? We get the following:

$$(a_1 - 1) + (a_2 - 2) + ... + (a_n - n) = (a_1 + a_2 + ... + a_n) - (1 + 2 + ... + n)$$
$$= (1 + 2 + ... + n) - (1 + 2 + ... + n)$$
$$= 0$$
$$= even.$$

Notice a couple things here. First, since the sequence $a_1, a_2, \ldots, a_n$ can be any rearrangement of the numbers $1, 2, \ldots,$ n, we have $a_1 + a_2 + \ldots + a_n = 1 + 2 + \ldots + n$ in some order. Second, zero is an even number because it is a (trivial) multiple of two.

Since we are assuming that n is odd, the sum $(a_1 - 1) + (a_2 - 2) + \ldots + (a_n - n)$ contains an *odd* number of terms. If every one of these n terms, $a_k - k$, were itself odd, then by Property 4 the sum would also have to be an *odd* number. But that contradicts the fact, as we saw previously, that the sum $(a_1 - 1) + (a_2 - 2) + \ldots + (a_n - n)$ is *even*. Therefore, at least one of the terms $a_k - k$ must be *even*. Then, by Properties 5 and 7, when we multiply all the terms $a_k - k$ to get $(a_1 - 1)(a_2 - 2)(a_3 - 3) \ldots (a_n - n)$, we conclude that the product $(a_1 - 1)(a_2 - 2)(a_3 - 3) \ldots (a_n - n)$ must be *even*. □

This problem was not too difficult, but all those "evens" and "odds" can make your head spin. To master problems like this, you must master the parity properties of the integers (Properties 1 through 7).

Problem 33

A Perfect Square

Prove that, for any positive integer n, $n(n+1)(n+2)(n+3)+1$ is a perfect square.

We are asked to prove that, for any positive integer n, we have $n(n+1)(n+2)(n+3)+1 = m^2$ for some positive integer m. Let's look at a few special cases to get familiar with the problem. For notational convenience, let $f(n) = n(n+1)(n+2)(n+3)+1$. If $n = 1$, we have $f(1) = 25 = 5^2$, which is a perfect square. If $n = 2$, we get $f(2) = 121 = 11^2$. And if $n = 3$, we get $f(3) = 361 = 19^2$. So, from the first three cases, it certainly seems as though $n(n+1)(n+2)(n+3)+1$ is a perfect square. How can we prove this?

There are many ways to prove the statement, and it is largely a matter of preference which method to use. We could, for example, prove the statement by mathematical induction. We could also prove it using modular arithmetic as we discussed in Problem 13. For example, we could use the fact that an odd square is congruent to 1 modulo 8. Another way to prove the statement, which is often used in mathematical problem solving, is the technique of *factoring*. We can start by expanding $f(n)$ as

$n(n+1)(n+2)(n+3)+1 = n^4 + 6n^3 + 11n^2 + 6n + 1$. Then we

factor the expression $n^4 + 6n^3 + 11n^2 + 6n + 1$. You could cheat

by using a good calculator that has a Computer Algebra System

(CAS). If you do that, you will get $n^4 + 6n^3 + 11n^2 + 6n + 1 =$

$(n^2 + 3n + 1)^2$. This is obviously a perfect square, so we are done.

That proves the problem statement. But how can we factor the

expression $n^4 + 6n^3 + 11n^2 + 6n + 1$ without cheating by using a

calculator?

A general method that is often used in solving many

mathematics problems is the *method of undetermined*

coefficients. Let's factor the expression $n^4 + 6n^3 + 11n^2 + 6n + 1$

using this method. Basically, we want to factor the expression as

a product of two identical polynomials (assuming it is possible)

as follows:

$$n^4 + 6n^3 + 11n^2 + 6n + 1 = (a_k n^k + a_{k-1} n^{k-1} + \ldots + a_0)^2.$$

Our task is now to determine the "undetermined coefficients"

$a_k, a_{k-1}, \ldots, a_0$. We can simplify the problem with a little

forethought. First, when we square the expression

$a_k n^k + a_{k-1} n^{k-1} + \ldots + a_0$, we must get a polynomial of degree four,

so we must have $k = 2$. Second, since the coefficient of n^4 is 1,

on the left-hand side of the above equation, we must have

$a_k = a_2 = 1$. Finally, we must have $a_0 = 1$, because the polynomial $n^4 + 6n^3 + 11n^2 + 6n + 1$ has a constant term equal to 1. With these simplifications, our problem is now to find the unknown coefficient a such that $n^4 + 6n^3 + 11n^2 + 6n + 1 = (n^2 + an + 1)^2$. Expanding the right-hand side, we get

$$n^4 + 6n^3 + 11n^2 + 6n + 1 = n^4 + 2an^3 + (a^2 + 2)n^2 + 2an + 1.$$

By comparing the coefficients of powers of n on both sides of this equation, we get $2a = 6$ and $a^2 + 2 = 11$. The value $a = 3$ certainly works. Plugging $a = 3$ into $(n^2 + an + 1)^2$ gives us our desired factorization:

$$n^4 + 6n^3 + 11n^2 + 6n + 1 = (n^2 + 3n + 1)^2. \ \square$$

The method of undetermined coefficients is a good technique to remember. It can be used whenever you can assume the *form* of a mathematical expression, and you just need to determine the unknown coefficients or parameters.

Problem 34

Ordered 4-Tuples

Let n be a positive integer. Find the number of ordered 4-tuples of integers (a, b, c, d) that satisfy $0 \le a \le b \le c \le d \le n$.

If n were, say, $n = 10$, then *one* solution would be (a, b, c, d) $= (2, 2, 5, 7)$, since $0 \le 2 \le 2 \le 5 \le 7 \le 10$. Clearly, there are many other solutions (a, b, c, d) that satisfy the condition $0 \le a \le b \le c \le d \le 10$. The problem does not ask us to actually find all of the solutions, and there could be many of them. Rather, the problem asks us to *count* how many solutions exist. This is a combinatorics problem.

It would be really nice if we could simply let (a, b, c, d) be any increasing ordered combination of four numbers chosen from the set $A = \{0, 1, 2, \ldots, n\}$. Then the answer would be the number of combinations of $n+1$ numbers, selected from the $n+1$ numbers in set A, chosen four at a time, or simply $\binom{n+1}{4} = \dfrac{(n+1)!}{4!(n-3)!}$. Here, you may remember that

$$\binom{n}{k} = \frac{n!}{k!(n-k)!}$$ is the number of combinations of n things

chosen k at a time (without repetition of selections).

Unfortunately, that approach won't work her, because some or all of the numbers a, b, c, d can be equal to each other. In other words, there can be repetitions of numbers in (a, b, c, d). For example, $(0, 0, 0, 0)$ is a solution. So, to make this idea work we would like to find a set of numbers B such that every combination of four *distinct* numbers $\{\alpha, \beta, \gamma, \delta\}$, ordered as $(\alpha, \beta, \gamma, \delta)$ with $\alpha < \beta < \gamma < \delta$, chosen from set B corresponds uniquely (bijectively) to a solution (a, b, c, d) of our original problem. Then, if $|B|$ is the size of set B, the answer to our problem will be $\binom{|B|}{4}$. How do we construct the set B?

In the original problem, (a, b, c, d) must satisfy $0 \le a \le b \le c \le d \le n$. Since $a \le b$, and both a and b are nonnegative integers, we must have $a < b+1$. So, we have $0 \le a < b+1$. Also, since $b \le c$, we must have $b < c+1$, and $b+1 < c+2$. So, $0 \le a < b+1 < c+2$. And since $c \le d$, we must have $c < d+1$, and $c+2 < d+3$. So, $0 \le a < b+1 < c+2 < d+3$. Finally, since $d \le n$, we have our final result:

$$0 \le a < b+1 < c+2 < d+3 \le n+3.$$

If we now let $\alpha = a$, $\beta = b+1$, $\gamma = c+2$, and $\delta = d+3$, we get

$$0 \le \alpha < \beta < \gamma < \delta \le n+3.$$

We have converted the original problem's inequality constraint from one using $\le$ into one using strict inequalities $<$. Now, if we select *any* four distinct numbers $(\alpha, \beta, \gamma, \delta)$, with $\alpha < \beta < \gamma < \delta$, from the set $B = \{0, 1, 2, \dots, n+3\}$, we will have a solution to our original problem. All we need to do, after selecting $(\alpha, \beta, \gamma, \delta)$ is convert it into (a, b, c, d) using the transformations $a = \alpha$, $b = \beta - 1$, $c = \gamma - 2$, and $d = \delta - 3$.

Since the set $B = \{0, 1, 2, \dots, n+3\}$ contains $n+4$ elements, or numbers, the solution to the original problem is $\binom{|B|}{4} = \binom{n+4}{4}$. In other words, the number of ordered 4-tuples of integers (a, b, c, d) that satisfy $0 \le a \le b \le c \le d \le n$ is $\binom{n+4}{4}$. $\square$

Problem 35

Logarithm Equation

Solve the following equation for real x:

$$\log_x 5 - 2\log_{5x} 5 - 4\log_{25x} 5 = 0.$$

We have a peculiar equation that involves logarithms, and we want to find x. It is possible that no real number x exists. It is also possible that x is not unique. There may be more than one real x that satisfies the given equation. These are fundamental questions—*existence* and *uniqueness*—that should be considered whenever we are asked to find some mathematical object. For now, let's just assume that there is a real x that satisfies the equation and see what happens.

If we are to successfully solve the equation for x, two things are certain. First, we will need to algebraically manipulate the equation to isolate x, though it is not immediately clear how to do that. Second, we will almost certainly need to use the properties of logarithms. So, let's start there. Let's review the properties of logarithms and see if there is anything that will help us solve our problem.

Here are several useful properties of logarithms. These are all *theorems*, but we will not prove them here.

1. For real $b > 0$, $\log_b 1 = 0$.

2. Identity Rule: For real $b > 0$, $\log_b b = 1$.

3. Addition Rule: For real $b > 0$, and positive reals m and n,
   $\log_b(mn) = \log_b m + \log_b n$.

4. Subtraction Rule: For real $b > 0$, and positive reals m and n, $\log_b(m/n) = \log_b m - \log_b n$.

5. Chain Rule: For positive reals a, b, c and $a \neq 1$, $b \neq 1$,
   $\log_a b \cdot \log_b c = \log_a c$.

6. Reciprocal Rule: For positive reals a, b and $a \neq 1$, $b \neq 1$,
   $$\log_a b = \frac{1}{\log_b a}.$$

7. Inverse Rule: For positive real $b \neq 1$, $\log_b(b^a) = a$.

8. Exponent Rule: For positive reals a, b, real n, and $b \neq 1$,
   $\log_b(a^n) = n\log_b a$.

If you carefully study these rules for logarithms, you might realize that the Reciprocal Rule is the tool we need for our problem. Why? The Reciprocal Rule exchanges, or interchanges, the base and the argument. In other words, the Reciprocal Rule converts $\log_a b$ into something that has $\log_b a$, namely $\dfrac{1}{\log_b a}$. So, the Reciprocal Rule will convert $\log_x 5$ into

$\dfrac{1}{\log_5 x}$. This is nice, because it gets the unknown value x out of

the base. It is much easier to work with $\log_5 x$ than $\log_x 5$.

Applying the Reciprocal Rule to each term of the equation $\log_x 5 - 2\log_{5x} 5 - 4\log_{25x} 5 = 0$, we get the following:

$$\frac{1}{\log_5 x} - \frac{2}{\log_5 5x} - \frac{4}{\log_5 25x} = 0.$$

This is a much nicer equation, because now all of the logarithms are to the same base, base 5. Now we just have to play around with some algebra.

Let $y = \log_5 x$. Then $\log_5(5x) = \log_5 5 + \log_5 x = 1 + y$. Also, $\log_5(25x) = \log_5 25 + \log_5 x = \log_5 5^2 + \log_5 x = 2\log_5 5 + \log_5 x = 2 + y$. Then, by substitution, we have

$$\frac{1}{y} - \frac{2}{1+y} - \frac{4}{2+y} = 0.$$

After simplification, the last expression becomes $5y^2 + 5y - 2 = 0$. This is a quadratic equation, and it can be solved using the quadratic formula (see Theorem 23.1 in Problem 23). When we solve for y, using the quadratic formula, we get $y = \dfrac{-5 + \sqrt{65}}{10}$ and $y = \dfrac{-5 - \sqrt{65}}{10}$. And since $y = \log_5 x$,

we have $x = 5^y$. This gives us two real solutions to the equation

$\log_x 5 - 2\log_{5x} 5 - 4\log_{25x} 5 = 0$:

$$x = 5^{(-5 + \sqrt{65})/10} \text{ and } x = 5^{(-5 - \sqrt{65})/10}. \ \square$$

Epilogue

A Tale of Mathematical Discovery

Our final lesson in the art of mathematical problem solving is a true story of how I discovered an interesting geometry theorem. The story shows us how the process of mathematical exploration and discovery really works, and it shows how you can also discover interesting new mathematical theorems.

Creativity rarely occurs in a vacuum, so, as is often the case in mathematics, we need some place to begin. We need a good problem to start working on. You might consider this problem to be a "seed" problem. Like a seed, we will grow it into a tree.

One day I was reading the book, *Mathematical Quickies*, by Charles W. Trigg (Appendix D, Recommended Reading). Problem number 201, entitled "Segments Determining an Equilateral Triangle," is a classic Euclidean geometry problem. Basically, we are given an equilateral triangle, as shown below, with a point P lying inside the triangle. It just so happens that when we connect the point P to each of the triangle's vertices, we obtain line segments of lengths 3, 4, and 5. The challenge is to find the length L, which is the side length of the equilateral triangle.

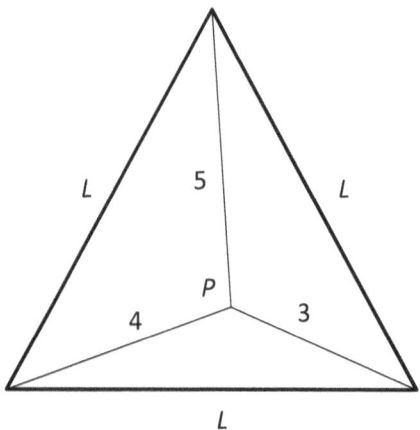

Note that this problem is asking us to find a particular solution for the special case where the distances from P to the vertices of the triangle are 3, 4, and 5. The book gives an ingenious solution using high school geometry. We won't repeat the solution here, because that is not our goal. The solution in the book takes advantage of the fact that there is something very special about the numbers 3, 4, and 5. Basically, there exists a Pythagorean right-angled triangle having sides of length 3, 4, and 5. And it is precisely that fact that allows this problem to be solved in a clever way. The solution turns out to be $L = \sqrt{25 + 12\sqrt{3}}$, or approximately 6.7664. But this is only the solution for the special case.

This is all very nice, but there is one problem. What happens if we pick an *arbitrary* point P, lying inside the equilateral triangle, and the distances from P to each of the triangle's vertices are a, b, and c—not necessarily 3, 4, and 5? How can we find the length L for this general problem? This is a common way by which mathematicians invent or discover new mathematics. First, we solve a particular problem. Then we investigate how to solve the *general problem*. The general problem looks like this, and we want to find L in terms of (as a function of) a, b, and c.

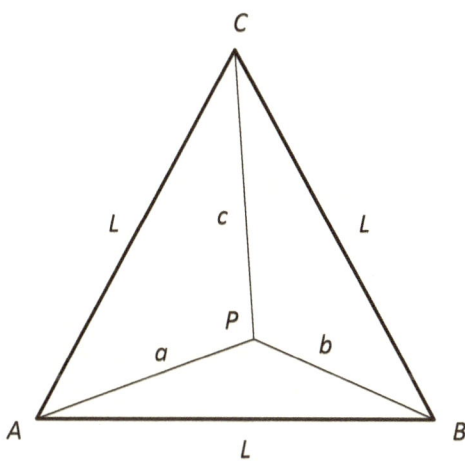

How can we solve this general problem? An obvious first approach is to try to repeat the textbook solution for the general case. Unfortunately, this doesn't work. The textbook solution took advantage of the special case that a, b, and c were 3, 4, and 5, which are the edge lengths of a special right-angled triangle:

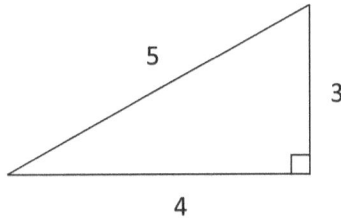

In general, lengths a, b, and c do *not* form a Pythagorean right-angled triangle. This means that the textbook solution will not work. We need to find another method.

A straightforward *brute-force-dumb* solution is the following. We observe that the area of triangle ABC is equal to the sum of the areas of triangles APC, CPB, and BPA:

$$Area(ABC) = Area(APC) + Area(CPB) + Area(BPA).$$

Since we want to find the length L purely in terms of the lengths a, b, and c, we can use Heron's formula for the areas of the triangles (Appendix B). This gives us the following nasty equation:

$$\sqrt{3}L^2 = \sqrt{(a+b+L)(-a+b+L)(a-b+L)(a+b-L)}$$
$$+ \sqrt{(a+c+L)(-a+c+L)(a-c+L)(a+c-L)}$$
$$+ \sqrt{(b+c+L)(-b+c+L)(b-c+L)(b+c-L)}$$

Now, in theory, we can solve the problem. We just have to solve this equation—involving lots of square roots—for the unknown quantity L. You can actually solve this ugly equation for L. If you are ferociously perseverant, and if you stay up late at night for nearly a week doing long, painful algebra calculations, you will eventually get the following solution:

$$L = \frac{\sqrt{a^2 + b^2 + c^2 + \sqrt{3}\sqrt{2a^2b^2 + 2a^2c^2 + 2b^2c^2 - a^4 - b^4 - c^4}}}{\sqrt{2}}.$$

If we have an equilateral triangle of side lengths L, and we pick an arbitrary point P *inside* the equilateral triangle, then this formula gives an expression for L in terms of the lengths $a = AP$, $b = BP$, and $c = CP$.

Now, I'm not saying that this is something that you really want to do. The brute-force-dumb algebraic method actually works—if you are willing to do a heroic amount of calculation.

Now that we know what the answer looks like, can we "polish the stone" and get something much nicer? This is where mathematics becomes a fine art. Like an artist, we want to look at the mathematical equation and see if we can find something more pleasing to look at. Here is one idea. The above formula for L is an *equality*. But if we discard some nonnegative terms from the right-hand side of the equality, we can turn it into an *inequality*.

Why do we need all that junk on the right-hand side that looks like $\sqrt{3}\sqrt{2a^2b^2 + 2a^2c^2 + 2b^2c^2 - a^4 - b^4 - c^4}$? This "junk" must be nonnegative, because we know that L is a nonnegative real number that actually exists. So, if we simply "drop," or remove, this from the right-hand side of the equation, we get the following beautiful inequality:

$$L \geq \sqrt{\frac{a^2 + b^2 + c^2}{2}} .$$

Let's test this inequality on the original problem where $(a, b, c) = (3, 4, 5)$. We have $L \geq \sqrt{\frac{3^2 + 4^2 + 5^2}{2}} = 5$. This is not too bad, since the solution to the original special case problem was $L = 6.7664$. In fact, the inequality $L \geq \sqrt{\frac{a^2 + b^2 + c^2}{2}}$ is a "best possible" inequality in the sense that sometimes we can actually achieve equality. This occurs, for example, if the point P, which lies inside the equilateral triangle ABC, actually coincides with point A (or B or C).

Let's briefly review what we have accomplished. We started with a textbook geometry problem that could be solved using special methods for a special case. Next, we asked how we could solve the general problem—not just the special case. To solve the general problem, we did an astronomically huge amount of

220

algebra to get an explicit formula for L in terms of a, b, and c. Then we "polished the stone" by removing "ugly" pieces of the formula to obtain the beautiful inequality $L \geq \sqrt{\dfrac{a^2 + b^2 + c^2}{2}}$.

Now that we know what this inequality looks like, we should ask ourselves whether we can prove it directly, using an elegant and simple proof.

Can we find a simple and elegant proof of this inequality? Where can we begin looking for a nice proof? When we look at the inequality, we notice that it has a sum of *squares*: $a^2 + b^2 + c^2$. Do we know of any geometry theorems that involve squares? We certainly do, namely the *Law of Cosines* (Appendix B, under Trigonometry).

Consider the following general equilateral triangle:

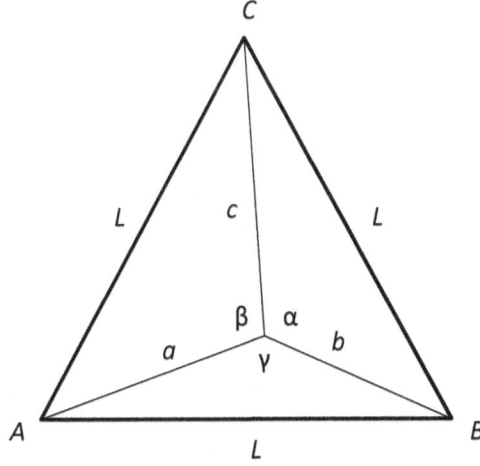

There are three triangles where we can apply the Law of Cosines to get three equations:

$$L^2 = a^2 + c^2 - 2ac\cos\beta$$
$$L^2 = a^2 + b^2 - 2ab\cos\gamma$$
$$L^2 = b^2 + c^2 - 2bc\cos\alpha$$

Since our inequality $L \geq \sqrt{\dfrac{a^2 + b^2 + c^2}{2}}$ involves a 2 in the

denominator, it seems highly likely that we only want to add *two* of the three cosine law equations. But which two? A key insight is that for point P being *inside* an equilateral triangle, at most one of the angles α, β, γ is acute (less than 90 degrees). That means that at least two angles are obtuse. Let's assume, for definiteness, that the obtuse angles are γ and α. Then, adding the two cosine laws $L^2 = a^2 + b^2 - 2ab\cos\gamma$ and $L^2 = b^2 + c^2 - 2bc\cos\alpha$ gives $2L^2 = a^2 + 2b^2 + c^2 - 2ab\cos\gamma - 2bc\cos\alpha$. Since, by hypothesis, angles α and γ are obtuse, we have $-2ab\cos\gamma \geq 0$ and $-2bc\cos\alpha \geq 0$. This means that we can drop these terms from the right-hand side of $2L^2 = a^2 + 2b^2 + c^2 - 2ab\cos\gamma - 2bc\cos\alpha$ to obtain the inequality $2L^2 \geq a^2 + 2b^2 + c^2$. We can now drop a b^2 term from the right-hand side to get $2L^2 \geq a^2 + b^2 + c^2$ and it immediately follows that $L \geq \sqrt{\dfrac{a^2 + b^2 + c^2}{2}}$.

Our first proof of this inequality was accomplished using brute-force-dumb calculations. After we discovered the inequality, we noticed that it should have a simpler derivation. By observing the *form* of the inequality, we were able to select some promising theorems and find an elegant proof of the inequality. This is how mathematical exploration and discovery usually work in the real world. Mathematicians do not sit down on a leisurely Sunday afternoon and casually produce great mathematical theorems. Great mathematical discoveries are most often the result of hard work. The secret is to find a good problem to work on, ask good basic questions about the problem, work really hard for a long time until you discover something interesting, and then "polish the stone." Turn your discoveries into beautiful works of art that are worthy of contemplation by the human mind.

I wish you the very best in your future mathematical explorations. You have everything that you need to discover great new mathematics—a good mind, paper, and pencil.

Appendix A

Problem Solving Dictums

Dictum 1. *Master the basics. And when you get lost in the forest, always go back to first principles.*

Dictum 2. *Even the most brilliant mathematicians have difficulties understanding mathematics.*

Dictum 3. *Fortune favors the bold (Ovid). Take bold risks.*

Dictum 4. *Embrace suffering. Failure is part of the process. The obstacle is the path (Zen proverb).*

Dictum 5. *Understand the problem. You cannot solve a problem that you don't understand.*

Dictum 6. *To solve a mathematics problem, you must explore the problem.*

Dictum 7. *Formally state things that are trivial and obvious. They are often the keys to solving your problem.*

Dictum 8. *Results are proportional to the amount of hard work you invest in a problem.*

Dictum 9. *In mathematics, genius is the result of hard work.*

Dictum 10. *Good problem solvers are flexible problem solvers.*

Dictum 11. *Identify your brute-force-dumb solution. How would you solve the problem if you were an idiot?*

Dictum 12. *It is easier to find the solution to a problem if you already know the answer.*

Dictum 13. *Look at small cases, special cases, and extreme cases.*

Dictum 14. *Try solving a more general problem that implies your problem as a special case.*

Dictum 15. *Exploit symmetry. If symmetry does not exist, see if you can artificially create it.*

Dictum 16. *See if your problem is equivalent, or similar, to another problem that you already know how to solve.*

Dictum 17. *Look for patterns and formulate conjectures.*

Dictum 18. *For problems in number theory and combinatorics, look for your numbers in Pascal's triangle.*

Dictum 19. *Don't try to be brilliant. Just explore the problem and see where that takes you.*

Dictum 20. *Let the problem suggest its own solution. (See Problem 31.)*

Dictum 21. *It is a mathematical certainty that you won't be able to solve every mathematics problem. That's OK.*

Dictum 22. *Practice, practice, practice.*

Dictum 23. *Ugly solutions usually precede beautiful solutions.*

Dictum 24. *Follow your passions and ignore the experts. Experts are usually right, but when they are wrong, they are spectacularly wrong.*

Appendix B

Useful Theorems

This appendix contains a handful of useful theorems that would be good to memorize. They are like tools in your toolbox. The list is by no means complete or comprehensive, but it's a good place to start. Knowing several good theorems will help you get a foothold on many problems by giving you a place to start your investigations and explorations.

Algebra

Theorem. $x(1-x) \leq \dfrac{1}{4}$, for all real x.

Theorem. $x + \dfrac{1}{x} \geq 2$, for all real $x > 0$.

Bernoulli's Inequality. For real $x > -1$, and natural number n, $(1+x)^n \geq 1 + nx$.

Triangle Inequality. For any two vectors $\mathbf{x}$ and $\mathbf{y}$ in R^n space, we have $|\mathbf{x} + \mathbf{y}| \leq |\mathbf{x}| + |\mathbf{y}|$, with equality if and only if $\mathbf{x}$ and $\mathbf{y}$

are proportional. The symbol "$|\mathbf{x}|$" denotes the magnitude of vector $\mathbf{x}$.

Cauchy's Arithmetic-Geometric Mean Inequality. For positive real numbers $x_1, x_2, \ldots, x_n$, $(x_1 + x_2 + \ldots + x_n)/n \geq (x_1 x_2 \ldots x_n)^{1/n}$, with equality if and only if all the x's are equal. This theorem states that the arithmetic mean is always greater than or equal to the geometric mean of n positive real numbers.

Definition - Harmonic Mean. If $x_1, x_2, \ldots, x_n$ are all positive real numbers, then the *harmonic mean* of the numbers is defined by $H = \dfrac{n}{\left(\dfrac{1}{x_1} + \dfrac{1}{x_2} + \ldots + \dfrac{1}{x_n}\right)}$. If A is the arithemetic mean, G is the geometric mean, and H is the harmonic mean, then $A \geq G \geq H$.

Theorem. Monotonic bounded sequences converge.

Rational Root Test. If $c_n x^n + c_{n-1} x^{n-1} + \ldots + c_1 x + c_0 = 0$ is a polynomial with integer coefficients and $x = a/b$ is a rational root in lowest terms, then b divides c_n and a divides c_0. (This theorem can often be used to prove that a number is irrational.)

Polynomial Remainder Theorem. When a polynomial $p(x)$ is divided by $(x - a)$, the remainder is $p(a)$.

Polynomial Factor Theorem (Corollary of Polynomial Remainder Theorem). The number a is a root of the polynomial $p(x)$ if and only if $x - a$ divides evenly into $p(x)$.

Identity Theorem for Polynomials. Let $p(x)$ and $q(x)$ be two polynomials in x over an infinite integral domain, each having degree less than or equal to n. If $p(x)$ and $q(x)$ have equal values for $n + 1$ or more distinct values of x, then the two polynomials are identical.

Descarte's Rule of Signs. Let $c_n x^n + c_{n-1} x^{n-1} + ... + c_1 x + c_0 = 0$ be a polynomial with real coefficients. Write down, in order, the signs (+ or −) of the nonzero coefficients. Then the number of positive roots is less than or equal to the number of sign changes. Example: Let $f(x) = 8x^9 - 7x^6 - 4x^5 + 3x^3 + 1$. The signs, in order, are +, −, −, +, +. The number of sign *changes* in this sequence is two (plus-to-minus, and minus-to-plus). Thus, the number of *positive* roots of $f(x) = 0$ is less than or equal to 2. So $f(x)$ may have 0, 1, or 2 positive roots.

Rational Equations. Let $p(x)$ and $q(x)$ be polynomials. The solution set S of the *rational equation* $p(x)/q(x) = 0$ is the set $S = \{x \mid p(x) = 0, q(x) \neq 0\}$.

Theorem. A polynomial of degree less than or equal to n is *uniquely* determined by specifying $n + 1$ of its values.

Huygen's Inequality. For $0 < x < \pi/2$, we have
$2\sin(x) + \tan(x) \geq 3x$.

Cauchy-Schwarz Inequality. For real numbers a_1, a_2, ... , a_n
and b_1, b_2, ... , b_n, we have $(a_1 b_1 + a_2 b_2 + ... + a_n b_n)^2 \leq$
$(a_1^2 + a_2^2 + ... + a_n^2)(b_1^2 + b_2^2 + ... + b_n^2)$, with equality if each a_j is
proportional to b_j. The square of the sum of products is less than
or equal to the product of the sums of squares.

Conjugate. The *conjugate* of $a + b\sqrt{d}$ is $a - b\sqrt{d}$. There is
also another definition of conjugate that applies to complex
numbers: The conjugate of $a + bi$ is $a - bi$.

Definition – Absolute Value. For real number x, the *absolute
value* of x is defined by $|x| = x$ if $x \geq 0$; $= -x$ if $x < 0$. Also,
$|x| = \sqrt{x^2}$.

Absolute Values. If x and y are real numbers, then

$$|x + y| \leq |x| + |y| \text{ (Triangle Inequality)},$$
$$|xy| = |x| \cdot |y|,$$
$$||x| - |y|| \leq |x - y|.$$

Factorization. $a^n - b^n = (a - b)(a^{n-1} + a^{n-2}b + ... + ab^{n-2} + b^{n-1})$.

Factorization. For n odd,

$$a^n + b^n = (a+b)(a^{n-1} - a^{n-2}b + \ldots - ab^{n-2} + b^{n-1}).$$

Definition - Convex Functions. A function $f(x)$ that is continuous and real-valued on an interval is called *convex* if, for any two points x and y in the interval, $f\left(\dfrac{x+y}{2}\right) \leq \dfrac{f(x)+f(y)}{2}$.

Jensen's Inequality. Let $f(x)$ be a real-valued function of a real variable x. Then $f(x)$ is convex if and only if for all u and v in the domain of $f(x)$ and for all t such that $0 \leq t \leq 1$, we have $f(tu + (1-t)v) \leq t \cdot f(u) + (1-t) \cdot f(v)$.

De Moivre's Theorem. For natural number n, $(\cos(x) + i \cdot \sin(x))^n = \cos(nx) + i \cdot \sin(nx)$, where $i = \sqrt{-1}$, the imaginary unit.

Trigonometry

Theorem. $-1 \le \sin(x) \le 1$, for all real x.

Theorem. $\sin(-x) = -\sin(x)$, for all real x.

Theorem. $\cos(-x) = \cos(x)$, for all real x.

Theorem. $\tan(x) = \dfrac{\sin(x)}{\cos(x)}$, provided $\cos(x) \ne 0$.

Theorem. $\tan(x+y) = \dfrac{\tan(x) + \tan(y)}{1 - \tan(x)\tan(y)}$, provided that $\tan(x)\tan(y) \ne 1$.

Theorem. $\sin(x + y) = \sin(x)\cos(y) + \cos(x)\sin(y)$.

Theorem. $\cos(x + y) = \cos(x)\cos(y) - \sin(x)\sin(y)$.

Theorem. $\sin^2(x) + \cos^2(x) = 1$.

Theorem. $1 + \tan^2(x) = \sec^2(x)$.

Theorem. $1 + \cot^2(x) = \csc^2(x)$.

Law of Cosines. For any triangle having sides a, b, and c, and angles α opposite side a, β opposite side b, and γ opposite side c, we have $a^2 = b^2 + c^2 - 2bc \cdot \cos(\alpha)$.

Extended Law of Sines. For any triangle having sides a, b, and c, and angles α opposite side a, β opposite side b, and γ opposite side c, we have $\dfrac{a}{\sin(\alpha)} = \dfrac{b}{\sin(\beta)} = \dfrac{c}{\sin(\gamma)}$. Furthermore, if the circumcircle of the triangle has radius R, then

$$\frac{a}{\sin(\alpha)} = \frac{b}{\sin(\beta)} = \frac{c}{\sin(\gamma)} = 2R.$$

Law of Tangents. For any triangle having sides a, b, and c, and angles α opposite side a, β opposite side b, and γ opposite side c we have $\dfrac{a+b}{a-b} = \dfrac{\tan\left((a+b)/2\right)}{\tan((a-b)/2)}$.

Number Theory

The Well-Ordering Principle. Every nonempty set of positive integers has a least (smallest) element.

The Division Algorithm. If a and b are integers with $b > 0$, then there exist *unique* integers q and r such that $a = qb + r$, with

$0 \le r < b.$

Bertrand's Postulate (Theorem). For integer $k > 1$, there exists a prime number between k and $2k$.

Legendre's Theorem. For prime p, the exponent of p in the prime factorization of $n!$ is $e_p(n!) = \sum_{r \ge 1} floor\left(\dfrac{n}{p^r}\right)$. The "floor" function $floor(x)$ gives the largest integer that does not exceed x. For example, $floor(\pi) = 3$, and $floor(-\pi) = -4$.

Euler's Totient Formula. If a positive integer n has the prime factorization $n = p_1{}^{\alpha} p_2{}^{\beta} \cdots p_m{}^{\mu}$, for distinct primes $p_1, p_2, \ldots, p_m$, then the number of integers between 1 and n, inclusive, that are coprime to n is given by $\varphi(n) = n \cdot \left(1 - \dfrac{1}{p_1}\right)\left(1 - \dfrac{1}{p_2}\right)\cdots\left(1 - \dfrac{1}{p_m}\right)$.

Corollary. If $n = p^r$, for prime p, then $\varphi(n) = p^r - p^{r-1}$.

Generalized Cancellation Law for Congruences. If $ka \equiv kb$ (mod n), then $a \equiv b$ (mod $n/(k, n)$), where (k, n) is the greatest common divisor of k and n.

Theorem. If k is the smallest positive integer for which $a^k \equiv 1$ (mod n), then k divides $\varphi(n)$, where $\varphi(n)$ is the Euler Totient function.

Fermat's Little Theorem. For prime p, and such that p does not divide integer a, $a^{p-1} \equiv 1$ (mod p).

Euler's Generalization of Fermat's Theorem. If integers a and m are coprime (i.e., $(a, m) = 1$, where (a, m) means "the greatest common divisor of"), then $a^{\varphi(m)} \equiv 1$ (mod m), where $\varphi(m)$ is the Euler Totient function.

Rule of "Casting Out Nines". Let $s(n)$ be the sum of the digits in the decimal representation of positive integer n. Then $n \equiv s(n) \bmod(9)$.

Divisibility by Eleven. Let $a_m a_{m-1} \ldots a_1 a_0$ be the decimal representation of a positive integer n. Then $n \equiv a_0 - a_1 + a_2 - a_3 + \ldots + (-1)^m a_m$ (mod 11). For example, 275 is divisible by 11 because $5 - 7 + 2 \equiv 0$ (mod 11).

Squares.

$\quad\quad n^2 \equiv 0$ (mod 4) if n is even,

$\quad\quad n^2 \equiv 1$ (mod 4) if n is odd,

$\quad\quad n^2 \equiv 0$ (mod 8) if $n \equiv 0$ (mod 4),

$n^2 \equiv 4 \pmod 8$ if $n \equiv 2 \pmod 4$,

$n^2 \equiv 1 \pmod 8$ if n is odd.

Lemma. If a and b are coprime integers (i.e., $(a, b) = 1$), then there exists an integer x such that $ax \equiv 1 \pmod b$.

Wilson's Theorem. For every prime p, $(p-1)! \equiv -1 \pmod p$. The converse is also true.

Quadratic Residues. An integer a that is coprime to an integer m is called a *quadratic residue* modulo m if there exists a solution of the congruence $x^2 \equiv a \pmod m$. Otherwise, a is a quadratic *non-residue* modulo m.

Legendre's Symbol. Let p be an odd prime (2 is the only even prime). Then *Legendre's symbol*, $\left(\dfrac{a}{p}\right)$ is defined by $\left(\dfrac{a}{p}\right) = 1$ if a is a quadratic residue modulo p; $= -1$ if a is a quadratic non-residue modulo p; $= 0$ if p divides a. Note that $\left(\dfrac{a}{p}\right)$ is just a notation; it does *not* mean "a divided by p."

Quadratic Reciprocity Law. Let p and q be distinct odd primes. Then

$$\left(\frac{p}{q}\right) = (-1)^{[(p-1)/2][(q-1)/2]}, \text{ where } \left(\frac{p}{q}\right) \text{ is Legendre's symbol.}$$

Chinese Remainder Theorem. If m_1, m_2, ..., m_k are *pairwise* coprime integers greater than 1, and a_1, a_2, ..., a_k are arbitrary integers, then there exists an integer x such that $x \equiv a_j \pmod{m_j}$ for all $j = 1, 2, ..., k$.

Theorem. $1 + 2 + 3 + ... + n = \dfrac{1}{2} n(n+1)$.

Theorem. $1^2 + 2^2 + 3^2 + ... + n^2 = \dfrac{1}{6} n(n+1)(2n+1)$.

Theorem. $1 + 3 + 5 + 7 + ... + (2n - 1) = n^2$. (The sum of the first n odd numbers is n^2.)

Binet's Formula for the Fibonacci Numbers. The Fibonacci sequence starts with 1, 1, then we add the numbers to get 2, then we add the last two numbers to get 3, add the last two numbers to get 5, and so on: 1, 1, 2, 3, 5, 8, 13, 21, 34, 55, ... Binet's formula is $F(n) = \dfrac{\alpha^n - \beta^n}{\sqrt{5}}$, where $\alpha = \dfrac{1+\sqrt{5}}{2}$ and $\beta = \dfrac{1-\sqrt{5}}{2}$.

Also, $\beta = \dfrac{-1}{\alpha}$. The roots of $x^2 - x - 1 = 0$ are α and β.

Recursive Formula for the Fibonacci Numbers. The Fibonacci numbers are defined recursively by

$F(n) = F(n-1) + F(n-2)$, for integer $n > 1$ and with initial conditions $F(0) = F(1) = 1$.

Pythagorean Number Triples. All primitive (integer) Pythagorean number triples (a, b, c) are given parametrically by $a = 2mn$, $b = m^2 - n^2$, and $c = m^2 + n^2$, where m and n are coprime, are of different parity (even/odd), and satisfy $m > n$. An example of a Pythagorean number triple is $(3, 4, 5)$. Note that these numbers satisfy the Pythagorean theorem for right-angled triangles: $3^2 + 4^2 = 5^2$.

The Möbius Function. $\mu(1) = 1$, $\mu(n) = (-1)^k$ if n is a product of k distinct primes, and $\mu(n) = 0$ otherwise.

The Möbius Inversion Formula. Let n be a positive integer. If $f(n) = \sum_{d|n} g(d)$, then $g(n) = \sum_{d|n} \mu(d) f(n/d)$.

Theorem. $\displaystyle\sum_{d|n} \mu(d) \frac{n}{d} = \varphi(n)$.

Theorem. $\displaystyle\sum_{d|n} \mu(d) = 0$, for integer $n \geq 2$.

240

Linear Diophantine Equations. Let a, b, and c be integers and let g be the greatest common divisor of a and b. Then the equation $ax + by = c$ has a solution in integers x and y if and only if g divides c. Also, if (x_0, y_0) is one integer solution of the equation $ax + by = c$, then a complete set of integer solutions is given by $\{(x_0 + bt/g, y_0 - at/g) \mid t \text{ is an integer}\}$.

An Inequality for Factorials. Let n be a positive integer. Then we have $n^{n/2} \leq n! \leq \dfrac{(n+1)^n}{2^n}$, where $n!$ is "n factorial." For example $5! = 5 \cdot 4 \cdot 3 \cdot 2 \cdot 1 = 120$.

Combinatorics

Theorem. The number of combinations of n distinct things chosen k at a time, without repetitions, is $\dbinom{n}{k} = \dfrac{n!}{k!(n-k)!}$.

Symmetry Condition for the Binomial Coefficients.

$$\binom{n}{k} = \binom{n}{n-k}.$$

Theorem. The number of combinations of n distinct things chosen k at a time, with repetitions allowed, is $\dbinom{n+k-1}{k}$.

Absorption/Extraction Identity. For $k \neq 0$,

$$\binom{n}{k} = \frac{n}{k}\binom{n-1}{k-1}.$$

Pascal's Identity. For $k > 0$, $\dbinom{n}{k} = \dbinom{n-1}{k} + \dbinom{n-1}{k-1}$.

Parallel Summation Identity.

$$\binom{n}{0} + \binom{n+1}{1} + \binom{n+2}{2} + \ldots + \binom{n+k}{k} = \binom{n+k+1}{k}$$

Sum of Products Identity.

$$\binom{r}{0}\binom{s}{n} + \binom{r}{1}\binom{s}{n-1} + \binom{r}{2}\binom{s}{n-2} + \ldots + \binom{r}{n}\binom{s}{0} = \binom{r+s}{n}$$

Trinomial Revision Identity.

$$\binom{n}{m}\binom{m}{k} = \binom{n}{k}\binom{n-k}{m-k}.$$

Upper Negation Identity. For integer k,

$$\binom{n}{k} = (-1)^k \binom{k-n-1}{k}.$$

Upper Summation Identity.

$$\binom{0}{m} + \binom{1}{m} + \binom{2}{m} + \dots + \binom{n}{m} = \binom{n+1}{m+1}.$$

Binomial Theorem. $(x+y)^n = \sum_{k=0}^{n} \binom{n}{k} x^k y^{n-k}$.

Newton's General Binomial Theorem.

$$\sum_{n=0}^{\infty} \binom{n+r-1}{n} x^n = (1-x)^{-r}, \text{ for } |x| < 1.$$

The Pigeonhole Principle. If $nk + 1$ pigeons, for integer $k \geq 1$, are placed in n pigeonholes, then at least one hole contains at least $k + 1$ pigeons.

K-gonal Numbers. The nth k-gonal number is given by

$$S_n^k = n + (k-2)\binom{n}{2}, \text{ where } \binom{n}{2} \text{ is the number of combinations of}$$

n things chosen 2 at a time. (The k-gonal numbers are numbers like triangular numbers, square numbers, pentagonal numbers, and so on.)

Catalan Numbers. The Catalan numbers are defined by

$$C(n) = \frac{1}{(n+1)}\binom{2n}{n}, \text{ for } n \geq 0. \text{ The Catalan sequence is 1, 1, 2, 5,}$$

14, 42, ...

Sperner's Lemma. Suppose that F is a family of subsets of the set $\{1, 2, \ldots, n\}$ such that no subset in the family contains another one. Then the size of F is bounded by $|F| \leq \binom{n}{\lfloor n/2 \rfloor}$, where $\lfloor n/2 \rfloor$ is the "floor" of n/2.

Topology and Analysis

Rolle's Theorem. Let $f(x)$ be a mapping from the closed interval $[a, b]$ to the reals: $f: [a, b] \to$ R. Also, let $f(x)$ be continuous on $[a, b]$ and differentiable on the open interval (a, b). If $f(a) = f(b)$, then there exists a number c in (a, b) such that $\dfrac{d}{dx} f(x)\big|_{x=c} = 0$.

The symbol $\dfrac{d}{dx}$ is the derivative operator.

Extreme Value Theorem. If $f(x)$ is continuous on the closed interval $[a, b]$, then there exist real numbers c and d in $[a, b]$ such that $f(c) \leq f(x) \leq f(d)$ for all x in $[a, b]$.

Intermediate Value Theorem. If $f(x)$ is continuous on the closed interval $[a, b]$ and if $f(a) < y < f(b)$ (or if $f(b) < y < f(a)$), then there exists a number c in the open interval (a, b) such that $f(c) = y$.

Mean-Value Theorem. Let $f(x)$ be a mapping from the closed interval $[a, b]$ to the reals: $f: [a, b] \to$ R. Also, let $f(x)$ be continuous on $[a, b]$ and differentiable on the open interval (a, b). Then there exists a number c in the interval (a, b) such that

$$\frac{d}{dx} f(x)\big|_{x=c} = \frac{f(b) - f(a)}{b - a}.$$

Bolzano's Theorem. Let $f(x)$ be continuous on the closed interval $[a, b]$. If $f(a)$ and $f(b)$ have opposite signs, then there exists a point c between a and b such that $f(c) = 0$.

Continuity of Polynomials. Polynomials are everywhere continuous, and rational functions $\dfrac{p(x)}{q(x)}$ are continuous at every point except where $q(x) = 0$.

Geometry

Theorem. The area of a triangle having height h and base b is $A = \dfrac{1}{2}bh$.

Theorem. Let T be a triangle having two of its side lengths equal to a and b. Let the angle between sides a and b be γ. Then the area of T is $A = \dfrac{1}{2}ab\sin\gamma$.

Heron's Formula. Let T be a triangle having side lengths a, b, and c. Let the semi-perimeter be $s = (a+b+c)/2$. Then the area of T is $A = \sqrt{s(s-a)(s-b)(s-c)}$.

Theorem. The line segment joining the midpoints of two sides of a triangle is parallel to the third side.

Cyclic Quadrilateral Theorem. A quadrilateral is cyclic if and only if each pair of opposite angles add up to 180 degrees (or π radians).

Subtended Angle Theorem. The angle subtended at the center of a circle by an arc is equal to double the angle subtended at the circumference.

Spherical Points Theorem. Given any four non-coplanar points, there exists a unique sphere which contains them.

Appendix C

Mathematical Tactics

There are many mathematical tactics that are frequently used to solve mathematics problems. Here is a reference list of some of the more common tactics.

☐ 1. Look at small cases.

Do some calculations using small numbers, or small values of parameters, to get familiar with the problem.

☐ 2. Look at special and extreme cases.

Examine special, extreme, or degenerate cases like zero, one, or infinity.

☐ 3. Do computations and generate data.

Do numerical computations and generate lots of data to discover patterns in the data.

☐ 4. Look at a simpler problem.

If the given problem is too hard to solve, try solving an easier, simpler, special case version of the problem.

□ 5. Look at a related problem.

If the given problem is too hard to solve, change or modify the problem to one that you can easily solve. Solving an easier, similar problem may give you insights on how to solve the harder problem.

□ 6. Solve a more general problem.

Solve a more general problem that implies your problem as a special case.

□ 7. Divide and conquer.

Break the problem into separate cases, then solve each case separately.

□ 8. Look for symmetry.

Look for symmetry in the problem that you can exploit. If the problem does not have obvious symmetry, try to create symmetry by adding auxiliary features or structure.

□ 9. Look for invariants.

An invariant is some feature of the problem that does not change. Parity (even or odd) is a common type of invariant. If the problem has an invariant, then choose an invariant configuration of the problem that is easy to solve.

☐ 10. Solve the dual problem.

Some problems have dual counterparts. In some cases the dual counterparts are equivalent. For example, fixing the area of a simple closed curve and minimizing its perimeter is equivalent to the dual problem of fixing the perimeter and maximizing the area.

☐ 11. Solve the complementary problem.

Some problems have complementary problems. In probability, for example, the probability that an event A occurs, or $P(A)$, is $1 - P(\sim A)$, where $P(\sim A)$ is the probability that event A does *not* occur.

☐ 12. Consider parity.

Consider "even" and "odd" cases separately. Is parity conserved or inavariant?

☐ 13. Order the objects or numbers.

Sort the objects or numbers into a logical order. Examine the extreme cases at the endpoints. Does the largest or smallest number have some special property?

☐ 14. Rearrange the objects or group them in pairs.

Try to rearrange the objects or numbers, or try to group them into special pairs based on some special property. Can you group the objects or numbers into invariant pairs?

☐ 15. Consider polar opposites.

Consider polar opposites such as "positive" and "negative" numbers. What happens if $x < 0$? What happens if $x > 0$?

☐ 16. Consider reciprocals.

See if you can transform the problem by considering reciprocals of some of the variables or parameters. For example, the reciprocal of x is $1/x$.

☐ 17. Factor polynomials.

Try to factor polynomials.

☐ 18. Change number bases.

Consider what happens if you change the number base. For example, what happens if you consider the numbers in base 2 (binary)?

☐ 19. Prime factorize positive integers.

Prime factorize the numbers in a problem to reveal potential hidden patterns.

☐ 20. Use inequalities.

Optimization problems can often be solved using inequalities such as the arithmetic-geometric mean inequality or the Cauchy-Schwartz inequality. If you cannot solve your problem exactly, inequalities can give you upper or lower bounds on the answer.

☐ 21. Average the data.

If individual data elements, or numbers, are not well-behaved, try considering averages of the data. The averages may be better behaved than the original data.

☐ 22. Look for your numbers in Pascal's triangle.

For combinatorial problems, look for your numbers in Pascal's triangle. This will often suggest a solution that involves binomial coefficients.

☐ 23. Use dimensional analysis.

Any physically correct solution to a real-world problem must be dimensionally correct. Dimensional analysis can often provide the correct form (up to a constant) of the solution.

☐ 24. Linearize nonlinear functions.

For nonlinear functions, consider their linear local approximations (e.g., by tangent line/plane approximations).

☐ 25. Remember algebraic identities.

Algebraic identities can often be used to solve difficult mathematics problems.

☐ 26. Use logarithms to chop down exponents.

Logarithms can be used to remove exponents. Logarithms can also be used to "chop" big or small numbers into manageable pieces, and logarithms can be used to convert multiplication problems into addition problems, and vice versa.

☐ 27. Try iteration and back-substitution.

Sometimes you can solve an equation either by back-substitution or by guessing a value and using repeated iteration in the given equation.

☐ 28. Consider one object as a "special" object.

Recurrence relations in combinatorics can often be derived by considering one object of a set to be a "special" object. Break the problem into two distinct cases—those that contain the special object and those that do not.

☐ 29. Form numerical ratios.

Form a dimensionless ratio of quantities that remains constant as the problem changes scale.

☐ 30. Look for similarity and proportionality.

Many geometry problems can be solved by identifying similar triangles or congruent triangles. For numerical problems, look for quantities that are proportional to each other.

☐ 31. Use double counting.

Combinatorial identities can often be proved by counting the same set of objects in two different ways. This is called *double counting.*

Appendix D

Recommended Reading

There are many excellent books about mathematical problem solving. Here are several books that I highly recommend. If you are new to the art of mathematical problem solving, then start with George Polya's classic, *How to Solve It.*

1. Barbeau, Klamkin, and Moser, *500 Mathematical Challenges*, The Mathematical Association of America, Washington, D.C., 1995.

2. Fomin, D., Genkin, S., and Itenberg, I., *Mathematical Circles (Russian Experience),* American Mathematical Society, Providence, Rhode Island, 1996.

3. Larson, L. C., *Problem-Solving Through Problems*, Springer-Verlag, New York, N.Y., 1983.

4. Lozansky, E., and Rousseau, C., *Winning Solutions*, Springer-Verlag, New York, N.Y., 1996.

5. Polya, G., *How to Solve It*, 2nd. ed., Princeton University Press, Princeton, N.J., 1957.

6. Polya, G., and Kilpatrick, J., *The Stanford Mathematics Problem Book*, Dover Publications, Inc., Mineola, New York, 2009.

7. Posamentier, A. S., and Lehmann, I., *Mathematical Curiosities*, Prometheus Books, Amherst, New York, 2014.

8. Trigg, C. W., *Mathematical Quickies,* Dover Publications, Inc., New York, N.Y., 1985.

9. Zeitz, P., *The Art and Craft of Problem Solving*, John Wiley & Sons, Inc., New York, N.Y., 1999.

Problem Sources

The problems presented in this book are from my collection. I have been collecting good mathematical problems from many sources over a period of several decades. The following list gives the sources of the problems to the best of my knowledge.

Problem 1. Posted on www.brilliant.org.

Problem 2. Mathematical Curiosities, Problem 84.

Problem 3. Posted on www.brilliant.org.

Problem 4. Mathematical Curiosities, Problem 30.

Problem 5. Quantum Magazine, June 1995.

Problem 6. Posted on www.brilliant.org.

Problem 7. Purdue University Problem of the Week.

Problem 8. Posted on www.brilliant.org.

Problem 9. Posted on www.brilliant.org.

Problem 10. Source unknown.

Problem 11. Posted on www.brilliant.org.

Problem 12. Stanford University Math Olympiad Problem Solving, Summer 2008.

Problem 13. Source unknown.

Problem 14. Posted on www.brilliant.org.

Problem 15. Posted on www.brilliant.org.

Problem 16. Source unknown. Mathematical classic.

Problem 17. Posted on www.brilliant.org.

Problem 18. Source unknown.

Problem 19. Posted on www.brilliant.org.

Problem 20. Posted on www.brilliant.org.

Problem 21. Source unknown.

Problem 22. 500 Mathematical Challenges, Problem 48.

Problem 23. Indian Mathematical Olympiad, date unknown.

Problem 24. Posted on www.brilliant.org.

Problem 25. Stanford University Math Olympiad Problem Solving, Summer 2008.

Problem 26. Posted on www.brilliant.org.

Problem 27. Stanford University Math Olympiad Problem Solving, Summer 2008.

Problem 28. Source unknown.

Problem 29. Source unknown.

Problem 30. Mathematical Quickies, Problem 263.

Problem 31. The Stanford Mathematics Problem Book, 2009.

Problem 32. Stanford University Math Olympiad Problem Solving, Summer 2008.

Problem 33. The Art and Craft of Problem Solving, p. 6.

Problem 34. Source unknown.

Problem 35. Canadian Mathematical Society.

Index

www.ingramcontent.com/pod-product-compliance
Lightning Source LLC
Chambersburg PA
CBHW031831170526
45157CB00001B/261